HOUSE OF ABUNDANCE PUBLICATIONS

The Abyss Inside

Mind-Blowing Facts and Discoveries About Your
Extraordinary Human Body

Understanding the human body is understanding the rhythm of life. Each beat of the heart, each breath, is a step in a dance choreographed by millions of years of evolution

Contents

1

Introduction

Welcome to "The Abyss Inside: Mind-Blowing Facts and Discoveries About Your Extraordinary Human Body." This is an exhilarating journey through the depths of human anatomy and physiology. Get ready to explore the intricate wonders within you as we unravel the mysteries and unveil the unique facts about your miraculous human body.

The Importance of Understanding the Human Body

Have you ever wondered what makes you who you are? With its awe-inspiring complexity, your body is not merely a vessel but a masterpiece of evolution. Understanding how this machinery functions is fascinating and essential for nurturing a healthier, more vibrant life.

By delving into the intricacies of the human body, we gain insights into how it operates, allowing us to make informed choices about our well-being. From unlocking the secrets of disease prevention to optimizing our physical and mental performance, this knowledge empowers us to take charge of our health.

This captivating book will explore the human body comprehensively, examining its various systems, organs, and functions. Each chapter will take you on a thrilling adventure through the network of bones, muscles, nerves, and organs that make up this astonishing vessel we call our body.

From the fascinating structures of the skeletal system to the pulsating rhythm of the circulatory system, from the workings of the nervous system to the complex dance of hormones in the endocrine system, we will uncover mind-blowing facts, discoveries, and intriguing insights that will leave you in awe of your existence.

Prepare to be amazed as we delve into the depths of human anatomy, discover the mysteries of bodily functions, and unveil the symphony of systems that work harmoniously to sustain your life. Along the way, we will explore the fascinating growth, development, and aging processes, shedding light on the profound changes that occur throughout the human lifespan.

Prepare to buckle up and gear yourself for an enthralling journey into "The Abyss Inside: Mind-Blowing Facts and Discoveries About Your Extraordinary Human Body." Get set to be spellbound, enlightened, and ignited by the breathtaking marvels within you.

2

Human Anatomy

Definition and Significance of Human Anatomy

Beneath the surface of our existence, a complex mosaic of components unfolds, each assigned a unique role and functionality. Step into the fascinating world of human anatomy as we examine the body's detailed architecture and comprehend its deep-seated importance.

Human anatomy is the study of the structure of the human body, examining the organization and arrangement of its various components. Through this exploration, we gain insights into how our bodies are constructed, from the microscopic building blocks to the grand design of interconnected systems.

Understanding human anatomy gives us a deeper appreciation for the incredible precision and complexity that allows our bodies to function seamlessly. It opens a gateway to knowledge about the body's diverse organs, tissues, and cells, unveiling their unique roles and contributions to our well-being.

But beyond its scientific importance, human anatomy holds profound significance. It allows us to connect with our bodies, fostering a sense of self-awareness and understanding. Through this exploration, we can truly marvel at the intricacies of our existence and develop a newfound respect for the vessel that carries us through life.

Overview of the Human Body's Structure and Organization

To fully understand the marvels of the human body, it's essential to comprehend its incredible structure and arrangement. In this portion, we'll set off on an illuminating voyage through the complexities of our physiological structure, revealing the astonishing framework that forms the foundation of our being.

The human body is a masterpiece of design, composed of various systems working harmoniously to sustain life. The encompassing array of systems comprises the skeletal system, muscular system, circulatory system, nervous system, respiratory system, digestive system, reproductive system, and numerous others. Each method has unique organs, tissues, and cells contributing to its specific functions.

The skeletal system is at the core of our physical framework, which provides support, protection, and mobility. This thorough network, composed of bones, cartilage, ligaments, and tendons, forms the foundation upon which our bodies are built. It gives us our shape and serves as a reservoir for essential minerals, produces blood cells, and houses the bone marrow.

The muscular system, collaborating with the skeletal system, allows us to make movements, from the most delicate gestures to the most energetic actions. Muscles, which are composed of fibrous tissues, contract and relax, thus generating the power necessary for various bodily movements. This intricate system, encompassing hundreds of muscles throughout our body, equips us with the capacity to walk, run, grip, and execute many other tasks.

The circulatory system facilitates the distribution of oxygen, nutrients, and waste products. It consists of the heart, blood vessels, and blood working together to ensure the transportation of vital substances throughout the body. The heart acts as a powerful pump, propelling oxygenated blood to tissues and organs while simultaneously receiving deoxygenated blood to be reoxygenated in the lungs.

The nervous system, comprising the brain, spinal cord, and a network of nerves, serves as the body's communication and control center. It enables us to perceive the world, process information, and coordinate our actions. From sensory perception to motor coordination, this system plays a fundamental role in our daily lives.

These are just a few glimpses into the remarkable organization and interplay of the human body's systems. In the upcoming chapters, we will delve deeper into each design, uncovering its functions, connections, and the fascinating discoveries that have expanded our understanding of ourselves.

Get ready to be astounded as we investigate the remarkable structure and arrangement of the human body. Collectively, let's decipher the enigmas beneath our skin and develop a deep admiration for the complex harmony of systems that constitute our identity.

Skeletal System and Its Role in Providing Support and Protection

The skeletal system serves as the foundation of our bodies, providing essential support, structure, and protection. Within its framework of bones lies a remarkable network that gives us our shape, safeguards our delicate internal organs, and facilitates movement.

The skeletal system, composed of bones, cartilage, ligaments, and tendons, is crucial in maintaining our overall physical integrity. Its primary functions extend beyond mere support; it serves as a mineral reservoir, produces blood cells, and facilitates bodily movements.

One of the key functions of the skeletal system is to provide structural support to the body. The bones form a sturdy framework that gives our bodies their shape and posture. They act as a scaffold for muscles, allowing them to attach and generate movements. Without the support of the skeletal system, our bodies would lack the stability required for everyday activities, such as standing, walking, and performing various tasks.

Moreover, the skeletal system serves as a protective shield for our internal organs. It encases vital structures, such as the brain, heart, lungs, and abdominal organs, shielding them from external impact and potential injury. The skull protects our brain, the ribcage safeguards our heart and lungs, and the vertebral column covers the delicate spinal cord.

In addition to support and protection, the skeletal system plays a critical role in locomotion. Bones, along with the muscles and joints, enable movement and provide leverage for muscle contraction. Through the coordinated action of bones and muscles, we can walk, run, jump, and perform detailed activities with precision and control.

Furthermore, the skeletal system contributes to the production of blood cells through a process called hematopoiesis. Within the bone marrow, specialized cells continuously generate new red and white blood cells, vital for oxygen transport, immune function, and overall health.

Muscular System and Its Function in Movement

The human body possesses a phenomenal ability to move and perform various physical activities, from the subtlest gestures to the most dynamic athletic feats. At the core of this remarkable capability lies the muscular system, a complex network of muscles that enables us to bend, flex, extend, and contract with precision and power.

The muscular system consists of three main types of muscles: skeletal, smooth, and cardiac. This section will focus primarily on skeletal muscles responsible for voluntary movement and play a fundamental role in our daily lives.

Skeletal or striated muscles are attached to bones by tendons and work in pairs to produce controlled movements. These muscles are composed of individual muscle fibers that contract and relax in response to nerve signals, resulting in coordinated and purposeful actions.

The primary function of the muscular system is to generate movement and maintain posture. When skeletal muscles contract, they pull on the bones, causing them to move. By working in synergy with the skeletal system, muscles allow us to perform an incredible array of activities, such as walking, running, lifting, and even sophisticated tasks like playing a musical instrument or typing on a keyboard.

Beyond movement, the muscular system also contributes to the stability and posture of our bodies. Specific muscles, known as postural muscles, work continuously to support the alignment of our spine and maintain an upright position against the force of gravity. These muscles are crucial in preventing postural imbalances and ensuring proper body alignment.

Additionally, the muscular system is responsible for generating heat. When muscles contract and relax, they produce heat as a byproduct of their metabolic processes. This heat production helps to regulate body temperature, allowing us to maintain a constant internal environment, even in fluctuating external conditions.

The muscular system also plays a vital role in protecting internal organs. Muscles encompass essential structures, such as the heart, composed of specialized cardiac muscles. The rhythmic contractions of cardiac muscles ensure continuous blood circulation throughout the body, supplying oxygen and nutrients while removing waste products.

Furthermore, the muscular system supports essential bodily functions beyond voluntary movement. Smooth muscles, found in the walls of organs, blood vessels, and the digestive system, enable involuntary movements, such as the rhythmic contractions of the digestive tract and the dilation and constriction of blood vessels.

Nervous System and Its Role in Communication and Coordination

The human body is a complex network of interconnecting systems, and at the helm of this sophisticated web lies the nervous system. With its remarkable ability to transmit signals and coordinate activities, the nervous system serves as the master conductor of the human orchestra.

The nervous system is responsible for communication, integration, and control throughout the body. It consists of two primary components: the central nervous system (CNS), which includes the brain and spinal cord, and the peripheral nervous system (PNS), which encompasses the network of nerves that extend throughout the body.

At the core of the nervous system is the brain, the epicenter of intelligence, consciousness, and thought. This awe-inspiring organ interprets sensory information, formulates responses, and orchestrates a symphony of commands that govern our actions and behaviors.

The spinal cord, a long, slender structure protected by the spinal column, is a vital conduit between the brain and the rest of the body. It relays signals to and from the brain, enabling coordinated movements, reflexes, and sensory experiences.

The peripheral nervous system extends its delicate tendrils outward from the CNS, reaching every corner of the body. This vast network includes sensory neurons that transmit information from sensory organs to the CNS and motor neurons that carry instructions from the CNS to muscles and glands.

The nervous system's fundamental unit is the neuron, a specialized cell that transmits electrical signals. Neurons communicate through synapses, microscopic junctions where signals are relayed from one neuron to the next.

Through this network, the nervous system ensures swift and precise communication, allowing us to perceive the world, react to stimuli, and maintain homeostasis. It enables us to experience sensations like touch, taste, smell, sight, and sound, shaping our perceptions and enriching our understanding of the environment.

But the nervous system's role goes far beyond sensory perception. It regulates and coordinates vital bodily functions, including heartbeat, respiration, digestion, and secretion. It governs the movement of our muscles, allowing us to walk, run, dance, and perform intricate tasks with astonishing agility.

The nervous system is also responsible for higher cognitive functions like memory, learning, reasoning, and emotions. It shapes our personalities, influences our behaviors, and enables the formation of deep connections with others.

As we venture into this exploration, we'll plunge into the fascinating aspects of the nervous system. Our journey will introduce us to its complex structures, such as neurons, glial cells, and many neurotransmitters facilitating intercellular communication. Delving further, we will decode the subtle mechanisms that govern sensory perception, motor control, and the delicate balance maintained by the autonomic nervous system's sympathetic and parasympathetic divisions.

3

Circulatory System

Introduction to the Circulatory System

Welcome to the awe-inspiring realm of the circulatory system, a remarkable network of vessels and organs that tirelessly transports life-giving substances throughout the human body. From the heart's rhythmic beating to the pathways of blood vessels, this chapter will unravel the mysteries of one of the most vital systems in our spectacular human bodies.

The circulatory system, also known as the cardiovascular system, is pivotal in maintaining the body's equilibrium and sustaining life. It is responsible for the circulation of blood, which acts as a transportation system, carrying essential nutrients, oxygen, hormones, and immune cells to every nook and cranny of our bodies.

At the heart of this remarkable system is the muscular organ that shares its name, the heart. The heart, nestled within the thoracic cavity, is a powerful pump that propels blood throughout the body with each rhythmic contraction. It is the driving force behind the continuous flow of blood, ensuring that vital substances reach every cell, tissue, and organ that depends on them.

The circulatory system has a complex blood vessel network that spans the entire body. Arteries, veins, and capillaries form an intricate web, carrying blood to and from various organs and tissues and facilitating the exchange of oxygen, nutrients, and waste products.

Embark with us on a captivating expedition through the boundless universe of the circulatory system. Witness how this sophisticated web flawlessly guarantees the delivery of life-essential elements to every nook and cranny of our bodies, offering the nourishment and oxygenation critical for peak functioning. Steel yourself to be astounded by this unique system's grand complexity and elegance, rhythmically throbbing to the beat of life. Together, let's decode the enigmas of the circulatory system, plunging into its profound depths and marveling at its remarkable capacity to sustain energy and foster vitality.

Components of the Circulatory System (Heart, Blood Vessels)

Within the multifaceted tapestry of the circulatory system, two key components reign supreme: the heart and the blood vessels. These remarkable structures work harmoniously to ensure the seamless flow of blood and the efficient distribution of life-sustaining substances throughout our bodies.

The Heart

At the epicenter of the circulatory system lies the heart, a marvel of muscular precision. This vital organ, roughly the size of a clenched fist, beats tirelessly, propelling blood through the vast network of blood vessels. It is truly the engine that keeps the circulatory system running smoothly.

The heart is divided into four chambers: two atria and two ventricles. The atria act as receiving chambers, while the ventricles function as the pumping powerhouses. These chambers work in perfect synchrony, allowing for the controlled and efficient circulation of blood.

The heart is equipped with valves to ensure unidirectional blood flow. These valves open and close with each heartbeat, allowing blood to move forward while preventing backflow. The coordinated contraction and relaxation of the heart's chambers, combined with the precise opening and closing of its valves, create the rhythmic symphony that drives circulation.

Blood Vessels

The circulatory system boasts a meticulous network of blood vessels that act as the conduits through which blood flows. These vessels come in three main types: arteries, veins, and capillaries, each playing a distinct role in the complex circulation web.

Arteries are the sturdy highways that carry oxygenated blood away from the heart and into the body's tissues. Their thick and elastic walls withstand the forceful contractions of the heart, ensuring that blood reaches even the farthest reaches of the body. Arteries gradually branch out into smaller vessels called arterioles, which further divide into the smallest vessels known as capillaries.

With their wafer-thin walls, capillaries form an interconnected network that connects arteries to veins. It is within these microscopically narrow passages that the exchange of oxygen, nutrients, and waste products occurs. Capillaries allow vital substances to diffuse out of the bloodstream and into the surrounding tissues, nourishing and facilitating their functions. Simultaneously, waste products, such as carbon dioxide, return to the capillaries to be carried away.

Veins, the return journey of the circulatory system, carry deoxygenated blood back to the heart. They gradually merge into larger vessels and eventually converge into two prominent veins, the superior and inferior vena cava, which return blood to the heart's right atrium. Veins have thinner walls than arteries and rely on valves to prevent the backward flow of blood, aiding its upward journey against gravity.

These components—heart, arteries, capillaries, and veins—collaborate in perfect synergy, creating a finely tuned system that ensures the delivery of oxygen, nutrients, hormones, and immune cells to every part of our bodies. Together, they navigate the vast landscape of the circulatory system, forging pathways of life and vitality.

Pulmonary and Systemic Circuits

Within the vast network of the circulatory system, two essential circuits work in tandem to ensure the proper functioning of our bodies: the pulmonary circuit and the systemic circuit. These circuits act as interconnected pathways, guiding blood flow to distinct destinations and serving different purposes.

Pulmonary Circuit

The pulmonary circuit, starting from the right side of the heart, involves transporting deoxygenated blood from various body parts. The blood enters the superior and inferior vena cava and is transferred to the lungs for oxygenation. Join us as we trace the path of this extraordinary circuit, observing the dramatic metamorphosis of blood as it undergoes its vital stages of transformation. From the right atrium, the blood flows into the right ventricle, which then contracts, propelling the blood through the pulmonary artery.

The pulmonary artery, the only artery in the body that carries deoxygenated blood, divides into smaller branches, subdivided into tiny capillaries that envelop the walls of the lung's air sacs, known as alveoli. Here, a remarkable exchange takes place. Carbon dioxide, a waste product of cellular respiration, diffuses out of the capillaries and into the alveoli. At the same time, oxygen from the inhaled air enters the capillaries and binds to hemoglobin within red blood cells.

Oxygenated blood then travels back to the heart, but this time, it enters the left side of the heart through the pulmonary veins. The left atrium receives the oxygenated blood and passes it into the left ventricle. With a powerful contraction, the left ventricle propels the oxygen-rich blood into the aorta, the largest artery in the body.

Systemic Circuit

The systemic circuit delivers oxygenated blood to all the organs, tissues, and cells throughout the body, nourishing them and providing them with the essential substances they need to thrive. Let us venture into this vast and detailed circuit, exploring the pathways it takes and the vital destinations it reaches.

The aorta emerges from the left ventricle, acting as the systemic circuit's main thoroughfare. The aorta branches out, forming smaller arteries that supply oxygenated blood to various body regions. These arteries further divide into arterioles, eventually leading to a compounded network of capillaries that permeate nearly every tissue and organ.

The exchange of oxygen, nutrients, and waste products occurs within the capillaries. Oxygen and nutrients diffuse out of the capillaries and into the surrounding tissues, providing nourishment and energy for cellular processes.

As blood continues its journey through the capillaries, it gradually converges into venules, which merge to form veins. These veins collect the deoxygenated blood and carry it back to the heart, specifically to the right atrium, initiating the cycle again.

Through the coordinated efforts of the pulmonary and systemic circuits, the circulatory system ensures the continuous flow of oxygenated blood, vital nutrients, and essential substances to every nook and cranny of our bodies. The pulmonary circuit facilitates the exchange of gases, replenishing oxygen supplies and eliminating carbon dioxide. In contrast, the systemic course nourishes our cells, enabling growth, repair, and optimal functioning.

The Function of Blood and Its Role in Transportation

Blood, the life-sustaining fluid coursing through our bodies, plays a vital role in transporting essential substances, ensuring the proper functioning of our organs and tissues. A meticulous symphony of cells and molecules collaborates within its crimson currents to deliver oxygen, nutrients, hormones, and waste products to their destinations.

Oxygen Transport

Transporting oxygen from the lungs to every cell in the body is one of the fundamental roles carried out by blood. Oxygen binds to hemoglobin, a protein found in red blood cells, forming a dynamic partnership for efficient oxygen delivery. As blood flows through the lungs during inhalation, oxygen molecules diffuse across the delicate alveolar-capillary membrane, binding to hemoglobin within red blood cells. As the circulatory system transports them, red blood cells carrying a load of oxygen embark on a remarkable journey to tissues and organs hungry for this vital element. Through this mechanism, blood ensures that cells receive the oxygen necessary for cellular respiration, which generates energy to fuel our bodily functions.

Nutrient Transport

Blood serves as a crucial conduit for delivering vital nutrients to our cells. From the nutrients derived from the foods we consume to the molecules synthesized by our organs, blood carries an array of sustenance to the hungry cells that depend on them. Through the systemic circuit, arteries, and capillaries transport nutrients such as glucose, amino acids, fatty acids, vitamins, and minerals to various tissues and organs. These nutrients are absorbed from the digestive system, extracted by the liver, and then released into the bloodstream to nourish cells. This intricate transport system ensures that our cells receive the building blocks and energy sources they need to sustain life.

Hormone Transport

Hormones, the body's chemical messengers, rely on blood to reach their target organs and elicit specific responses. Endocrine glands, such as the pituitary, thyroid, adrenal glands, and others, release hormones into the bloodstream, where they hitch a ride to their intended destinations. Blood acts as the courier, transporting hormones to distant organs and tissues, binding to specific receptors, and initiating essential physiological processes. Whether regulating metabolism, growth, reproduction, or maintaining homeostasis, blood facilitates the timely delivery of hormones, ensuring the smooth coordination of countless bodily functions.

Waste Product Removal

Just as blood transports vital substances, it also plays a crucial role in removing waste products from our cells and tissues. Metabolic byproducts, such as carbon dioxide and other waste molecules, accumulate as cells carry out their activities. Blood serves as a scavenger, collecting these waste products and carrying them away to organs responsible for their elimination. Carbon dioxide, a waste product of cellular respiration, is transported by the blood from the tissues to the lungs, where it is exhaled. Similarly, other waste products are transported to the kidneys, liver, and other organs for processing and removal from the body. Through this process, blood maintains the delicate balance necessary for optimal health.

The multifaceted nature of blood and its role as a transportation system is nothing short of astonishing. It serves as a lifeline, nourishing our cells, removing waste, and facilitating essential communication within our bodies.

The complicated interplay of red blood cells, plasma, and various other components ensures the seamless delivery of oxygen, nutrients, hormones, and waste to their destinations.

Common Diseases and Conditions Related to the Circulatory System

While the circulatory system is a remarkable network that sustains our bodies, it is not impervious to ailments. Various diseases and conditions can affect this vital system's delicate balance and functioning. Understanding these common disorders is crucial for recognizing their symptoms, seeking appropriate medical care, and adopting preventive measures. Let us explore some of the most prevalent diseases and conditions related to the circulatory system.

Hypertension (High Blood Pressure)

Hypertension is a condition characterized by persistently elevated blood pressure levels. It places excessive strain on the blood vessels, heart, and other organs, increasing the risk of serious complications such as heart disease, stroke, and kidney problems. Hypertension often has no symptoms but can be detected through regular blood pressure measurements. Lifestyle modifications, medication, and managing underlying health conditions are commonly employed to control hypertension and minimize its impact on the circulatory system.

Coronary Artery Disease (CAD)

Coronary artery disease occurs when plaque, consisting of cholesterol and other substances, builds up in the arteries supplying the heart with blood. Over time, this can narrow or block the arteries, restricting blood flow to the heart muscle. CAD can lead to angina (chest pain), heart attacks, and heart failure. Lifestyle changes, medications, and medical procedures such as angioplasty or bypass surgery are often employed to manage CAD and restore blood flow to the heart.

Stroke

A stroke occurs when the blood supply to a part of the brain is interrupted or reduced, leading to brain cell damage and potentially permanent neurological deficits. Ischemic stroke, caused by a blockage in a blood vessel supplying the brain, and hemorrhagic stroke, caused by bleeding in the brain, are the two main types. Prompt medical attention is crucial in managing strokes to minimize damage and maximize recovery. Rehabilitation and preventive measures are essential in reducing the risk of recurrent strokes.

Peripheral Artery Disease (PAD)

Peripheral artery disease affects the arteries outside the heart and brain, typically those supplying the limbs. It is characterized by the narrowing or blockage of these arteries, leading to reduced blood flow to the legs and arms. Symptoms may include leg pain, cramping, and slow wound healing. Lifestyle changes, medication, and medical procedures are employed to manage PAD and improve blood circulation to the affected limbs.

Deep Vein Thrombosis (DVT)

Deep vein thrombosis occurs when a blood clot forms in one of the deep veins, commonly in the legs. DVT can be caused by prolonged immobility, surgery, or certain medical conditions. Failure to receive treatment may result in the dislodgment of the clot, which can then migrate to the lungs, leading to a critical condition known as pulmonary embolism, which poses a potential threat to life. Medication, compression stockings, and lifestyle modifications are employed to prevent and treat DVT.

Atherosclerosis

Atherosclerosis is a condition characterized by the buildup of plaque inside the arteries. This plaque consists of cholesterol, fatty substances, calcium, and other materials that narrow and harden the arteries, restricting blood flow to vital organs. Atherosclerosis can affect various arteries in the body and significantly contribute to heart disease, stroke, and peripheral artery disease. Lifestyle changes, medication, and medical procedures may be employed to manage atherosclerosis and prevent complications.

These are just a few diseases and conditions that can affect the circulatory system. Maintaining a healthy lifestyle, managing risk factors such as high blood pressure and cholesterol levels, and seeking timely medical care are essential for preserving the health of this system. Understanding these common disorders can protect our circulatory system and ensure optimal functioning, allowing us to lead vibrant and fulfilling lives.

4

Digestive System

The digestive system is a complex network of organs and processes that enables our bodies to break down food, absorb essential nutrients, and eliminate waste products. It is vital in providing the fuel and building blocks necessary for our growth, development, and overall well-being. Let's embark on the fascinating world of the digestive system and explore its essential components and functions.

The digestive system begins in the mouth, where food is initially ingested and broken down into smaller particles through chewing and mixing with saliva. The food then travels down the esophagus, a muscular tube that connects the mouth to the stomach, through peristalsis—a series of coordinated contractions that propel the food forward.

Upon reaching the stomach, the food encounters gastric juices and enzymes, which help break it down further and facilitate the digestion of proteins. The stomach's muscular walls churn and mix the food, forming a partially digested semi-liquid substance called chyme. This chyme is gradually released into the small intestine, where most nutrient absorption occurs.

The small intestine, a long and intricately folded tube, consists of three segments: the duodenum, jejunum, and ileum. Here, the breakdown of carbohydrates, proteins, and fats reaches its completion with the help of various digestive enzymes produced by the pancreas and intestinal cells. Nutrients, such as amino acids, glucose, and fatty acids, are then absorbed into the bloodstream and transported to the body's cells for energy production, growth, and repair.

The remaining undigested material, including fiber, passes into the large intestine, also known as the colon. In the colon, water is reabsorbed, and the waste products are formed into solid feces. The feces are stored in the rectum until eliminated through the anus during a bowel movement.

Throughout this complex process, the digestive system relies on the coordinated actions of various organs, including the liver, gallbladder, and pancreas, which play crucial roles in producing and delivering digestive enzymes, bile, and other substances necessary for proper digestion.

Understanding the functioning of the digestive system helps us appreciate the incredible complexity and efficiency with which our bodies process the food we consume. By optimizing our dietary choices, maintaining a balanced lifestyle, and addressing any digestive concerns, we can support the healthy functioning of our digestive system and promote overall wellness.

Organs Involved in Digestion

The digestive system relies on a series of organs working harmoniously to break down food and extract nutrients. Let's explore the critical organs involved in the process of digestion.

Mouth

The journey of digestion begins in the mouth, where food is ingested, and initial mechanical breakdown occurs through chewing. The saliva secreted by salivary glands helps moisten the food, making it easier to swallow and initiating the chemical analysis of starches with the enzyme amylase.

Esophagus

Once the food is swallowed, it travels down the esophagus, a muscular tube connecting the mouth to the stomach. Rhythmic contractions propel the food downward through peristalsis, ensuring its safe passage to the stomach.

Stomach

The stomach is a muscular organ that acts as a reservoir for food and aids in further mechanical and chemical digestion. It secretes gastric juices containing hydrochloric acid and enzymes, such as pepsin, which break down proteins. The stomach's muscular contractions mix the food with gastric juices, forming chyme. This partially digested mixture is gradually released into the small intestine.

Small Intestine

The small intestine is a long, coiled tube where most nutrient absorption occurs. It consists of three segments: the duodenum, jejunum, and ileum. The small intestine receives digestive enzymes from the pancreas and bile from the liver and gallbladder to aid in the breakdown of carbohydrates, proteins, and fats. The small intestine's inner lining is covered in finger-like projections called villi, which significantly increase the surface area for nutrient absorption into the bloodstream.

Large Intestine

As the remaining undigested and unabsorbed food reaches the large intestine, water and electrolytes are reabsorbed, forming feces. The large intestine also houses beneficial bacteria that aid in the breakdown of undigested fibers and produce specific vitamins.

These organs work in a coordinated manner to break down food into smaller molecules, ensuring optimal absorption and utilization of nutrients by the body. Understanding the roles of these organs in the digestive system provides us with a deeper appreciation for the intricacies involved in turning food into usable energy and building blocks for our bodies.

Role of Accessory Organs (Liver, Pancreas, Gallbladder)

In addition to the primary organs involved in digestion, the digestive system relies on several accessory organs that play crucial roles in the process. These organs work together to ensure the efficient breakdown and absorption of nutrients. Let's dive into the remarkable functions of the liver, pancreas, and gallbladder.

Liver

The liver is the largest internal organ in the human body and performs multiple functions vital to digestion. One of its primary roles is the production of bile, a greenish-yellow fluid that aids in the digestion and absorption of fats. Bile is stored and concentrated in the gallbladder before being released into the small intestine. The liver also metabolizes nutrients, detoxifies harmful substances, stores vitamins and minerals, and produces essential proteins for blood clotting and immune function.

Pancreas

The pancreas is an endocrine and exocrine gland that releases hormones directly into the bloodstream and digestive enzymes into the small intestine. The pancreatic enzymes, including amylase, lipase, and proteases, are crucial for breaking down carbohydrates, fats, and proteins. These enzymes are released into the small intestine to aid food digestion further. Additionally, the pancreas produces insulin and glucagon, hormones regulating blood sugar levels.

Gallbladder

The gallbladder is a small, pear-shaped organ located beneath the liver. Its primary function is to store and concentrate bile produced by the liver. When fatty foods enter the small intestine, the gallbladder contracts, releasing bile into the digestive tract. Bile helps emulsify fats, breaking them into smaller droplets and facilitating their digestion by enzymes.

The coordinated actions of the liver, pancreas, and gallbladder are crucial for optimal digestion and nutrient absorption. These accessory organs work in harmony with the primary organs to ensure the effective breakdown of food and the absorption of essential nutrients into the bloodstream.

Common Digestive Disorders and Their Impact on Health

The digestive system is a complex network of organs and processes that combine to break down food and extract nutrients. However, this system can sometimes be susceptible to various disorders that significantly impact an individual's health and well-being. Let's explore some common digestive disorders and their effects on the body.

Gastroesophageal Reflux Disease (GERD)

GERD is characterized by the backward flow of stomach acid into the esophagus. This can cause symptoms such as heartburn, chest pain, regurgitation, and difficulty swallowing. Over time, untreated GERD can lead to complications like esophageal inflammation, ulcers, and an increased risk of esophageal cancer.

Peptic Ulcer Disease

Peptic ulcers are open sores that develop on the stomach lining, small intestine, or esophagus. These ulcers are often caused by a bacterial infection called Helicobacter pylori or prolonged nonsteroidal anti-inflammatory drugs (NSAIDs) use. Symptoms may include abdominal pain, bloating, nausea, and vomiting. Peptic ulcers can lead to severe complications such as bleeding, perforation, or obstruction without proper treatment.

Inflammatory Bowel Disease (IBD)

IBD is an umbrella term for chronic conditions that cause inflammation in the digestive tract. The two primary forms of IBD are Crohn's disease and ulcerative colitis. Both conditions can lead to symptoms such as abdominal pain, diarrhea, rectal bleeding, fatigue, and weight loss. IBD is a lifelong condition that requires ongoing management and can significantly impact a person's quality of life.

Irritable Bowel Syndrome (IBS)

IBS is a functional digestive system disorder characterized by symptoms like abdominal pain, bloating, constipation, and diarrhea. While the exact cause of IBS is unknown, triggers may include certain foods, stress, and hormonal changes. Although IBS doesn't cause permanent damage to the digestive tract, it can greatly affect a person's daily life and well-being.

Celiac Disease

Celiac disease is an autoimmune condition activated upon ingesting gluten, a protein prevalent in wheat, barley, and rye. When individuals with celiac disease consume gluten, their immune system responds by damaging the lining of the small intestine, leading to various symptoms such as abdominal pain, diarrhea, fatigue, and nutrient deficiencies. Adherence to a rigorous gluten-free diet is the sole treatment for celiac disease.

These are just a few examples of the many digestive disorders that individuals may encounter. Understanding these conditions and their impact on health is crucial for early detection, proper management, and overall well-being.

5

Respiratory System

T he respiratory system is a remarkable network of organs and tissues that enables us to breathe, providing our bodies with the essential oxygen needed for survival. From the moment we take our first breath to the countless inhalations and exhalations throughout our lives, the respiratory system plays a vital role in sustaining our existence. In this chapter, we will embark on a fascinating journey into the workings of the respiratory system.

Importance of the Respiratory System

The respiratory system exchanges oxygen and carbon dioxide between our bodies and the surrounding environment. It ensures that oxygen is taken in and transported to every cell in the body while simultaneously removing carbon dioxide, a waste product of cellular metabolism. Without this continuous gas exchange process, our bodies would be deprived of the oxygen necessary for energy production. They would be overwhelmed by the buildup of toxic carbon dioxide.

Structure and Components

The respiratory system has several key components, each with a specific role in facilitating respiration. It includes the nose, nasal cavity, pharynx, larynx, trachea, bronchi, bronchioles, and lungs. These structures work together seamlessly to carry out the processes of ventilation, gas exchange, and oxygen utilization within our bodies.

Mechanism of Breathing

Breathing, or pulmonary ventilation, is the process by which air is drawn into the lungs and expelled from the body. It involves the coordinated actions of the diaphragm, intercostal muscles, and various respiratory muscles. Understanding the mechanics of breathing and the factors that influence respiratory rate and depth is essential for comprehending the functioning of the respiratory system.

Gas Exchange and Transport

The exchange of gases, primarily oxygen, and carbon dioxide, occurs within the tiny air sacs called alveoli, which are found in the lungs. Here, oxygen from the inhaled air diffuses into the bloodstream. At the same time, carbon dioxide, produced as a waste product by our cells, moves from the blood into the alveoli to be exhaled. The respiratory system works with the circulatory system to efficiently transport gases to and from tissues throughout the body.

Respiratory Regulation

The respiratory system is tightly regulated to maintain homeostasis and ensure a constant supply of oxygen and carbon dioxide removal. Various factors, including neural mechanisms, chemical sensors, and feedback loops, influence the control of breathing. These regulatory processes help adjust our breathing rate and depth in response to changing body oxygen and carbon dioxide levels.

As we immerse ourselves in the details of the respiratory system, we are bound to gain a deep admiration for its indispensable role in the continuation of human life. The forthcoming chapters will bring us a closer look at each part's structure and function, decode the processes of gas exchange and transport, and scrutinize prevalent respiratory disorders and their influence on our health. United in our pursuit, we will reveal the astounding intricacies of the respiratory system, thereby deepening our comprehension of its fundamental importance in our day-to-day existence.

Structures Involved in Respiration

Within the intricate web of the respiratory system, several key structures work harmoniously to facilitate the process of respiration.

The Nose

The nose serves as the primary entrance for air into the respiratory system. Its external portion, the nostrils, allows the intake of oxygen-rich air. As air enters the nasal cavity, it is filtered, humidified, and warmed by the mucous membranes and tiny hair-like structures called cilia. These mechanisms help protect the delicate respiratory system from foreign particles and ensure that the air we breathe is clean and suitable for the subsequent respiration processes.

The Trachea

The trachea, also known as the windpipe, is a tube-like structure that connects the larynx to the bronchi. It consists of cartilaginous rings that provide structural support, preventing airway collapse during respiration. Within the trachea, specialized cells called goblet cells produce mucus, which helps to trap foreign particles and facilitate the removal of debris from the respiratory system.

The Lungs

The lungs are the central organs of the respiratory system, responsible for the exchange of gases between the air and the bloodstream. Protected by the rib cage, the lungs are two spongy, cone-shaped structures in the thoracic cavity. They comprise millions of tiny air sacs called alveoli, where oxygen and carbon dioxide exchange occurs. A double-layered membrane surrounds the lungs called the pleura, which helps to reduce friction during breathing.

Bronchi and Bronchioles

Within the lungs, the trachea branches into two main bronchi, one leading to each lung. These bronchi further divide into smaller bronchioles, which extend deep into the lung tissue. Bronchioles are lined with smooth muscles that regulate airflow and play a crucial role in maintaining proper ventilation.

Alveoli

The alveoli are the microscopic air sacs where the exchange of gases occurs. These thin-walled structures provide an extensive surface area for the diffusion of oxygen from inhaled air into the bloodstream and the release of carbon dioxide into the exhaled air. The delicate alveolar walls are richly supplied with tiny blood vessels called capillaries, ensuring efficient gas exchange.

Mechanics of Breathing and Gas Exchange

The respiratory system's mechanics of breathing and gas exchange are fascinating processes that allow our bodies to constantly supply oxygen to our cells and remove carbon dioxide, maintaining a delicate balance vital for survival. Let's investigate the mechanisms involved in this remarkable journey.

When we inhale, the process begins with the contraction of the diaphragm, a dome-shaped muscle located at the base of the chest. Simultaneously, the intercostal muscles between the ribs contract, causing the ribcage to expand. These actions create negative pressure within the chest cavity, drawing air into the lungs through the airways.

Common Respiratory Conditions and Their Effects on Breathing

With its delicate balance of structures and processes, the respiratory system can be susceptible to various conditions that impact our breathing. Let's explore some of the most common respiratory disorders and their effects on the respiratory system.

Asthma

Asthma is a chronic condition characterized by inflammation and narrowing of the airways. This leads to recurring wheezing, shortness of breath, chest tightness, and coughing. During an asthma attack, the muscles surrounding the airways contract, causing further narrowing and making breathing difficult. Asthma can be triggered by various factors such as allergens, exercise, or respiratory infections.

Chronic Obstructive Pulmonary Disease (COPD)

COPD is a progressive lung disease that encompasses chronic bronchitis and emphysema. Chronic bronchitis involves inflammation and excessive mucus production in the airways, leading to coughing and difficulty breathing. Emphysema is characterized by damage to the alveoli, reducing the lungs' ability to exchange oxygen and carbon dioxide efficiently. COPD is often caused by long-term exposure to cigarette smoke or other irritants.

Pneumonia

Pneumonia is characterized by the inflammation of air sacs in either one or both lungs due to an infection. Bacteria, viruses, or fungi are potential culprits that can induce pneumonia. Pneumonia can lead to coughing, fever, chest pain, and difficulty breathing. In severe cases, it can significantly impair lung function and require medical intervention.

Allergic Rhinitis

Allergic rhinitis, known as hay fever, is an allergic reaction triggered by airborne substances such as pollen, dust mites, or pet dander. It causes inflammation of the nasal passages, leading to symptoms like sneezing, itching, nasal congestion, and a runny nose. These symptoms can affect breathing and overall respiratory comfort.

Sleep Apnea

Sleep apnea is a sleep disorder characterized by pauses in breathing or shallow breaths during sleep. It occurs when the muscles in the throat fail to keep the airway open, leading to brief interruptions in breathing. Sleep apnea can result in loud snoring, fragmented sleep, excessive daytime sleepiness, and reduced oxygen levels during sleep.

These are just a few respiratory conditions that can impact breathing and overall respiratory health. Understanding these conditions and their effects is essential for early detection, proper management, and respiratory well-being.

6

Nervous System

T he nervous system is an intricate network of cells and tissues that plays a fundamental role in communication and coordination throughout the body. It is responsible for processing sensory information, initiating and regulating body movements, and maintaining homeostasis.

The nervous system can be divided into two primary components: the central nervous system (CNS) and the peripheral nervous system (PNS). The CNS consists of the brain and the spinal cord, which serve as the command center for the body's activities. The PNS includes the nerves that extend from the CNS to various body parts, enabling communication between the CNS and the rest of the body.

Within the nervous system, neurons and glial cells transmit and process information. Neurons are specialized cells that transmit electrical impulses, allowing for the rapid communication necessary for coordinated body functions. Glial cells provide support and protection to neurons, maintaining their environment and assisting in their functioning.

The nervous system can be further categorized into different divisions based on its functions. The somatic nervous system controls voluntary movements and relays sensory information, such as touch and temperature, to the CNS. In contrast, the autonomic nervous system regulates involuntary heart rate, digestion, and breathing. The autonomic nervous system is further divided into the sympathetic and parasympathetic divisions, which work harmoniously to maintain balance in response to internal and external stimuli.

In addition to its vital role in controlling bodily functions, the nervous system also encompasses the special senses of vision, hearing, taste, and smell. The eyes, ears, tongue, and nose serve as sensory receptors, gathering information about the external environment and relaying it to the brain for interpretation.

Structure and Function of Neurons and Glial Cells

Neurons and glial cells are the essential building blocks of the nervous system, each playing a unique role in transmitting and processing information. Let's explore the structure and function of these remarkable cells.

Neurons, which are referred to as nerve cells, form the bedrock of the nervous system. They are specialized to transmit electrical impulses, or nerve impulses, to communicate and coordinate various functions within the body. Structurally, neurons have three main components: the cell body, dendrites, and axons.

The cell body, called the soma, contains the nucleus and other essential organelles. It serves as the control center of the neuron, responsible for maintaining cell functions and synthesizing essential proteins necessary for its operation.

Dendrites are branching extensions that receive incoming signals from other neurons or sensory receptors. These signals, in the form of electrical impulses, travel across the dendrites and converge at the cell body. Dendrites are crucial in integrating and relaying information to the cell body.

Axons are long, slender projections that carry nerve impulses from the cell body to other neurons or target cells, such as muscles or glands. Covered by a protective myelin sheath, axons facilitate the rapid transmission of electrical signals over long distances. The myelin sheath is formed by specialized glial cells known as oligodendrocytes in the CNS and Schwann cells in the PNS.

At the end of axons are specialized structures called axon or synaptic terminals. These terminals form connections, called synapses, with other neurons or target cells. Within the synapse, electrical impulses are converted into chemical signals that can then be transmitted to the next neuron or target cell.

Glial cells, or neuroglia, provide crucial support and protection to neurons. Although often overlooked, these cells are essential for properly functioning the nervous system. Glial cells outnumber neurons and come in several types, including astrocytes, oligodendrocytes, microglia, and ependymal cells.

Astrocytes are star-shaped cells that help maintain the chemical environment surrounding neurons. They regulate the concentration of ions and neurotransmitters, provide nutrients to neurons, and play a role in repairing damaged brain tissue.

Oligodendrocytes and Schwann cells produce the myelin sheath, insulating axons and enhancing the speed of nerve impulse transmission. In addition to their role in myelination, oligodendrocytes provide structural support to neurons in the CNS.

Microglia function as the immune cells of the nervous system. They survey the brain for pathogens or abnormal cells. They are activated in response to injury or infection, aiding tissue repair and regulating inflammation.

Ependymal cells line the fluid-filled spaces within the brain and spinal cord. They contribute to the production and circulation of cerebrospinal fluid, which provides cushioning and support to the nervous system.

Neurons and glial cells work together in a coordinated manner, allowing for the complex functions of the nervous system. Neurons transmit electrical impulses, while glial cells provide support and maintenance. This multi-faceted partnership ensures the proper functioning of the nervous system and enables the transmission of information throughout the body.

Central Nervous System: The Brain and Spinal Cord

The central nervous system (CNS) comprises two vital components: the brain and the spinal cord.

The brain, often considered the body's command center, is a complex organ responsible for various functions, including cognition, emotions, memory, and voluntary movements. Structurally, the brain consists of distinct regions with unique roles and responsibilities.

The cerebral cortex, the brain's outer layer, is critical in higher cognitive functions, such as perception, language, reasoning, and decision-making. It is divided into two hemispheres—left and right—interconnected by a bundle of nerve fibers called the corpus callosum.

Beneath the cerebral cortex lie deep structures known as the basal ganglia, which are involved in coordinating movement and regulating voluntary motor control. These structures work with the cerebral cortex and other regions to facilitate smooth and purposeful actions.

The limbic system, situated deep within the brain, involves emotions, memory formation, and motivation. It includes the amygdala, hippocampus, and hypothalamus, which regulate emotional responses, form memories, and control essential physiological functions like hunger, thirst, and sleep.

The brainstem, connecting the brain to the spinal cord, controls vital functions necessary for survival, such as breathing, heart rate, and blood pressure. It also serves as a relay center, transmitting sensory and motor information between the brain and the rest of the body.

Moving on to the spinal cord is a long, cylindrical bundle of nerve fibers extending from the base of the brain down through the spinal column. The spinal cord is a vital conduit for communication between the brain and the peripheral nervous system (PNS).

Within the spinal cord, sensory neurons transmit information from the body's periphery to the brain, allowing us to perceive sensations such as touch, pain, and temperature. Motor neurons, on the other hand, carry signals from the brain to the muscles and glands, enabling voluntary movements and regulating physiological processes.

The brain and spinal cord form a complex and interconnected network, facilitating communication and coordination throughout the body. These organs work with the peripheral nervous system to process sensory information, initiate appropriate responses, and maintain homeostasis.

In the subsequent sections, we will explore the subtle functions of different brain regions, the remarkable capabilities of the spinal cord, and how they collectively contribute to our perception, behavior, and overall well-being. The central nervous system represents the pinnacle of evolutionary development, allowing us to navigate the world and experience the wonders of human consciousness.

Peripheral Nervous System and Its Subdivisions

Beyond the central nervous system, the human body houses an interconnected network known as the peripheral nervous system (PNS). The PNS extends throughout the body, connecting the central nervous system to various organs, muscles, and sensory receptors. Let's review the subdivisions of the peripheral nervous system and its essential functions.

Somatic Nervous System (SNS)

The somatic nervous system is responsible for voluntary movements and sensory perception. It enables us to interact with the external environment and consciously control our actions. Motor neurons within the SNS transmit signals from the central nervous system to skeletal muscles, initiating purposeful movements. On the other hand, sensory neurons carry information from sensory organs and receptors to the central nervous system, providing us with sensations like touch, temperature, and pain.

Autonomic Nervous System (ANS)

The autonomic nervous system operates involuntarily, regulating vital functions necessary for sustaining life. It controls the activities of internal organs, glands, and smooth muscles. The ANS consists of two main subdivisions: the sympathetic nervous system and the parasympathetic nervous system.

Sympathetic Nervous System

The sympathetic nervous system is responsible for the body's "fight-or-flight" response. It prepares the body for action in response to perceived threats or stressors. Activating the sympathetic nervous system leads to increased heart rate, elevated blood pressure, dilated pupils, and the release of stress hormones like adrenaline, mobilizing the body's resources for immediate action.

Parasympathetic Nervous System

In contrast to the sympathetic system, the parasympathetic nervous system promotes relaxation and restoration. It counterbalances the effects of the sympathetic system, facilitating rest, digestion, and recovery. Activation of the parasympathetic system slows the heart rate, enhances digestion, and conserves energy.

The coordination between the sympathetic and parasympathetic systems ensures a delicate balance in maintaining homeostasis, adapting to different situations, and responding appropriately to internal and external stimuli.

Enteric Nervous System (ENS)

The enteric nervous system is a unique division of the peripheral nervous system that resides within the walls of the gastrointestinal tract. It functions independently and controls digestion, nutrient absorption, and food movement through the intestines. The ENS contains an extensive network of neurons and is crucial in regulating gastrointestinal functions, such as peristalsis, secretion, and blood flow within the digestive system.

The peripheral nervous system, including the somatic, autonomic, and enteric divisions, works in harmony with the central nervous system to ensure the proper functioning of the entire body. It enables us to interact with the world, maintain physiological balance, and adapt to changing environments.

Role of the Nervous System in Sensory Perception and Motor Control

The nervous system, encompassing the central and peripheral components, plays a crucial role in our ability to perceive and interact with the world. A complex network of neurons and processes enables sensory perception and motor control, allowing us to experience sensations and initiate purposeful movements. Let's uncover the remarkable functions of the nervous system in sensory perception and motor control.

Sensory Perception

Sensory perception is how we receive and interpret information from our environment through various sensory modalities. The nervous system enables us to perceive and make sense of sensations such as touch, sight, hearing, taste, and smell. Here's an overview of how the nervous system contributes to sensory perception:

Sensory Receptors

Specialized sensory receptors throughout the body detect stimuli from the environment or within the body. These receptors convert different forms of energy, such as light, sound, or pressure, into electrical signals called nerve impulses.

Sensory Neurons

Nerve impulses generated by sensory receptors travel along sensory neurons, which transmit the signals to the central nervous system for processing. Each sensory modality has dedicated pathways that relay information to specific brain regions, where it is interpreted and integrated into meaningful perceptions.

Sensory Processing

In the brain, sensory information undergoes complex processing and interpretation. Different brain regions analyze and combine sensory inputs, allowing us to recognize objects, perceive depth, differentiate between colors, and interpret the qualities of various stimuli. This meticulous process ultimately forms our perception of the world.

Motor Control

Motor control refers to the ability to initiate and regulate voluntary movements. The nervous system is fundamental in coordinating and executing precise motor actions.

Motor Cortex

The motor cortex, located in the brain's frontal lobe, is responsible for planning, initiating, and controlling voluntary movements. It sends signals to the appropriate muscles through the spinal cord, enabling purposeful actions.

Motor Neurons

Motor neurons in the spinal cord and brainstem transmit signals from the motor cortex to the muscles. These signals, known as motor commands, trigger muscle contractions and coordinated movements.

Neuromuscular Junction

At the point where motor neurons connect with muscle fibers, known as the neuromuscular junction, neurotransmitters are released, initiating muscle contraction. This process allows for precise control of movements and the execution of complex actions.

Feedback Mechanisms

During movement, the nervous system continuously receives sensory feedback from muscles, joints, and other body parts. This feedback lets the brain monitor and adjust motor commands, ensuring coordinated and accurate movements.

The intricate interplay between sensory perception and motor control showcases the remarkable capabilities of the nervous system. By integrating sensory information and translating it into appropriate motor responses, the nervous system enables us to navigate our environment, interact with objects, and engage in activities essential to our daily lives.

Neurological Disorders and Their Impact on the Body

Neurological disorders are conditions that affect the structure or function of the nervous system, resulting in disruptions to normal brain, spinal cord, and peripheral nerve activity. These disorders can profoundly affect the body's functioning, leading to many symptoms and impairments. Let's explore some common neurological disorders and their impact on the body.

Alzheimer's Disease

Alzheimer's is a progressive neurodegenerative disorder characterized by memory loss, cognitive decline, and behavioral changes. It primarily affects the brain, causing a gradual deterioration of brain cells and the formation of abnormal protein structures. As the disease progresses, individuals may experience difficulties with memory, thinking, language, and daily activities.

Parkinson's Disease

Parkinson's disease is a chronic neurodegenerative disorder that affects movement. It occurs due to the loss of dopamine-producing cells in the brain, leading to motor symptoms such as tremors, rigidity, bradykinesia (slowness of movement), and postural instability. Non-motor symptoms, including depression, sleep disturbances, and cognitive impairment, can also manifest.

Multiple Sclerosis

Multiple sclerosis (MS) is characterized by an autoimmune response targeting the central nervous system. In MS, the immune system mistakenly attacks the protective covering of nerve fibers, disrupting communication between the brain and the rest of the body. This results in many symptoms, including fatigue, muscle weakness, coordination problems, sensory disturbances, and cognitive impairments.

Epilepsy

Epilepsy is a neurological disorder characterized by recurrent seizures. Seizures occur due to abnormal electrical activity in the brain, leading to temporary disruptions in normal brain function. The symptoms and severity of seizures can vary widely, ranging from momentary lapses of awareness to convulsive movements and loss of consciousness.

Stroke

A stroke occurs when the blood supply is disrupted to the brain, leading to brain cell damage. The effects of a stroke depend on the area of the brain affected and the extent of the damage. Common symptoms include sudden weakness or paralysis on one side of the body, speech difficulties, vision problems, and cognitive impairments.

Migraine

Migraine, a neurological disorder, manifests as recurring headaches and additional symptoms like nausea, sensitivity to light and sound, and visual disturbances. Migraines can significantly impact a person's quality of life. Various factors, including hormonal changes, certain foods, stress, or environmental factors may trigger them.

Amyotrophic Lateral Sclerosis (ALS)

ALS, commonly called Lou Gehrig's disease, is a progressive neurodegenerative condition that impacts the nerve cells in the brain and spinal cord. It leads to the degeneration of motor neurons, resulting in muscle weakness, loss of coordination, and eventually paralysis. ALS affects voluntary muscle control, impacting speech, swallowing, and mobility.

These are just a few examples of the numerous neurological disorders, each with unique characteristics and impact on the body. Neurological disorders can affect various aspects of physical and cognitive functioning, leading to mobility, communication, cognition, and overall well-being challenges.

Understanding these disorders and their effects is crucial for developing effective treatments, improving patient care, and enhancing the quality of life for individuals with neurological conditions. Ongoing research and advancements in medical science offer hope for improved management and potential breakthroughs in neurological disorders.

7

Reproductive System

The reproductive system is a fascinating and essential aspect of the human body, responsible for creating new life and continuing the human species. It encompasses a complex network of organs, hormones, and processes that work together to facilitate reproduction. In this chapter, we're set to explore the fascinating aspects of the reproductive system, diving deep into its exceptional functionalities.

The reproductive system serves two primary purposes: the production of gametes (sperm and eggs) and the provision of a supportive environment for developing and nurturing offspring. It is divided into two distinct systems: the male reproductive system and the female reproductive system, each with its unique structures and functions.

Throughout this chapter, we will examine the reproductive anatomy, hormonal regulation, and the intricacies of the reproductive processes in both males and females. We will explore the development of reproductive organs, the production of gametes, and the journey of fertilization leading to the creation of new life.

Additionally, we will discuss the various factors that influence reproductive health, such as hormonal balance, fertility, and contraception. Understanding these factors is crucial for individuals and couples planning to start a family or seeking to maintain their reproductive well-being.

We will also address common reproductive disorders and conditions impacting fertility and reproductive function. By shedding light on these conditions, we aim to provide insights into their causes, symptoms, and available treatments, offering a comprehensive understanding of reproductive health issues.

The reproductive system is vital for the continuation of the human species. It is deeply intertwined with personal identity, emotions, and relationships. We will explore the psychological and social aspects of reproduction, including the impact of reproductive choices and the importance of reproductive education and healthcare.

Female Reproductive Anatomy and Physiology

The complex network of organs within the female reproductive system collaboratively enables the reproductive process and nurtures the growth of new life. Understanding the female reproductive anatomy and physiology is crucial for appreciating the remarkable capabilities of the female body.

Internal and external organs can be found within the female reproductive system. Externally, the primary structures include the mons pubis, labia majora, labia minora, clitoris, and vaginal opening. The mons pubis is a fatty tissue pad covering the pubic bone. The labia majora and labia minora are folds of skin that protect the vaginal and urethral openings. The clitoris, located at the top of the labia minora, is a sensitive organ involved in sexual pleasure. The vaginal opening serves as the entrance to the reproductive tract.

Internally, the female reproductive system comprises several key organs. The main organ is the uterus, also known as the womb, responsible for nurturing and supporting the developing embryo and fetus during pregnancy. The uterus is connected to the ovaries by the fallopian tubes, which transport eggs from the ovaries to the uterus. The ovaries on each side of the uterus produce eggs (ova) and release hormones, including estrogen and progesterone, which regulate the menstrual cycle and support pregnancy.

The cervix is the lower, narrow portion of the uterus that connects it to the vagina. During childbirth, the cervix dilates to allow the baby's passage from the uterus into the birth canal. The vagina is a muscular tube that serves as the birth canal and is also the site for sexual intercourse. It connects the cervix to the external genitalia.

The female reproductive system undergoes cyclic changes known as the menstrual cycle. The menstrual cycle involves the release of an egg from one of the ovaries, known as ovulation, and the preparation of the uterus for the possible implantation of a fertilized egg. If fertilization does not occur, the lining of the uterus is shed during menstruation, marking the start of a new menstrual cycle.

Hormones play a vital role in regulating the female reproductive system. The hypothalamus in the brain releases gonadotropin-releasing hormone (GnRH), which stimulates the pituitary gland to produce follicle-stimulating hormone (FSH) and luteinizing hormone (LH). FSH stimulates the growth and maturation of follicles in the ovaries, each containing an egg. LH triggers ovulation and stimulates the production of estrogen and progesterone.

The female reproductive system is a phenomenal and sophisticated entity, capable of fostering the miracle of new life. Comprehending its structure and functioning is essential for preserving women's health, ensuring fertility, and thoughtful family planning. By understanding the subtleties within the female reproductive system, we can cultivate a profound respect for its awe-inspiring potential and the significant part it performs in the wonder of human reproduction.

Male Reproductive Anatomy and Physiology

The male reproductive system is a complex network of organs and structures that work together to produce, store, and deliver sperm for reproduction. Understanding the male reproductive anatomy and physiology is critical to comprehending the processes involved in male fertility and reproduction.

The male reproductive system consists of both internal and external organs. Externally, the primary structures include the penis and the scrotum. The penis serves as the organ of copulation and delivers sperm into the female reproductive tract during sexual intercourse. The scrotum, a pouch of skin below the penis, houses the testicles and helps regulate the temperature required for optimal sperm production.

Internally, the male reproductive system comprises the testes, epididymis, vas deferens, seminal vesicles, prostate gland, and bulbourethral glands. The testes within the scrotum produce sperm and testosterone, the primary male sex hormone. The epididymis, a coiled tube connected to each testicle, acts as a storage and maturation site for sperm.

The vas deferens is a muscular tube that carries mature sperm from the epididymis to the seminal vesicles, which mixes with seminal fluid to form semen. The seminal vesicles contribute fructose-rich fluid that provides energy for the sperm. The prostate gland produces a milky fluid that enhances sperm motility and neutralizes acidity in the female reproductive tract. The bulbourethral glands secrete a clear lubricating fluid that aids in the passage of semen during ejaculation.

A complex interplay of hormones regulates the male reproductive system. The hypothalamus in the brain releases gonadotropin-releasing hormone (GnRH), which stimulates the pituitary gland to produce follicle-stimulating hormone (FSH) and luteinizing hormone (LH). FSH promotes sperm production (spermatogenesis) in the testes, while LH stimulates testosterone production.

Spermatogenesis is a continuous process in which immature sperm cells undergo several stages of development to become mature, motile sperm. The mature sperm are then released into the epididymis for storage until ejaculation occurs.

During sexual arousal, the male reproductive system undergoes physiological changes. These changes include the engorgement of blood vessels in the penis, resulting in an erection, and the contraction of various muscles to propel semen through the urethra during ejaculation.

Understanding the male reproductive anatomy and physiology is essential for comprehending fertility issues, reproductive health, and preventing and treating male reproductive disorders. By gaining insights into the intricacies of the male reproductive system, we can develop a deeper appreciation for its remarkable capabilities and crucial role in human reproduction.

Puberty, Menstruation, and Fertility

Puberty marks a significant milestone in both males and females, representing the onset of sexual maturation and the potential for reproduction. It is a transformative period characterized by physical, hormonal, and emotional changes. During puberty, the male and female reproductive systems undergo remarkable development, leading to distinct processes related to fertility.

Puberty typically begins between 9 and 14 in both sexes, although the exact timing can vary among individuals. It is triggered by the release of hormones from the hypothalamus in the brain, specifically gonadotropin-releasing hormone (GnRH). GnRH signals the pituitary gland to release follicle-stimulating hormone (FSH) and luteinizing hormone (LH), stimulating the testes in males and the ovaries in females to produce sex hormones.

One of the most noticeable changes during puberty is the development of secondary sexual characteristics. In males, these include the growth of facial and body hair, deepening of the voice, and increased muscle mass. In females, the changes involve the development of breasts, the widening of hips, the appearance of pubic and underarm hair, and the onset of menstruation.

Menstruation, or the menstrual cycle, is a natural process exclusive to females in which the lining of the uterus is shed through the vagina. It is a monthly occurrence that typically lasts around 28 days, although cycle lengths can vary. The menstrual cycle is regulated by the fluctuation of hormones, primarily estrogen and progesterone.

During each menstrual cycle, an egg is released from one of the ovaries in ovulation. If the released egg is fertilized by sperm, it may implant in the uterus, resulting in pregnancy. If fertilization does not occur, the lining of the uterus is shed, resulting in menstrual bleeding. This process does not appear in males.

Fertility, or the ability to conceive and bear children, typically begins after puberty in both males and females. It is important to note that fertility varies among individuals and can be influenced by various factors, including hormonal balance, overall health, and lifestyle choices. Understanding one's reproductive system and fertility patterns can empower family planning and reproductive health.

Puberty, menstruation, and fertility are significant aspects of the male and female reproductive journeys. They represent the biological mechanisms that enable individuals to conceive and nurture new life. By understanding and embracing these processes, individuals can make informed decisions about their reproductive health, contraception, and family planning, ensuring the well-being of themselves and their future generations.

Common Reproductive Health Issues and Treatments

With its detailed and delicate balance, the reproductive system is susceptible to various health issues affecting both males and females. These reproductive health concerns require attention, diagnosis, and appropriate treatment to ensure optimal well-being. Here, we explore some common reproductive health issues and available treatments.

Infertility

Infertility is when a person or couple cannot conceive a child despite regular and unprotected sexual intercourse. It can affect both males and females and may result from various factors such as hormonal imbalances, structural abnormalities, or underlying medical conditions. Infertility treatments include fertility medications, assisted reproductive technologies (ART) like in vitro fertilization (IVF), and surgical interventions.

Sexually Transmitted Infections (STIs)

STIs are infections transmitted through sexual contact and can cause significant reproductive health problems. Common STIs include chlamydia, gonorrhea, syphilis, genital herpes, and human papillomavirus (HPV). Depending on the specific infection, treatment for STIs typically involves antibiotics, antiviral medications, or other targeted therapies.

Polycystic Ovary Syndrome (PCOS)

PCOS is a hormonal disorder affecting females, characterized by enlarged ovaries with small cysts, hormonal imbalances, and irregular menstrual cycles. It can lead to fertility issues, weight gain, acne, and excessive hair growth. Management of PCOS may involve lifestyle changes, hormone therapy, and medications to regulate menstrual cycles and address associated symptoms.

Benign Prostatic Hyperplasia (BPH)

BPH is a non-cancerous prostate gland enlargement that commonly occurs in aging males. It can cause urinary problems, such as frequent urination and weak urine flow. Treatment options for BPH include:

- Medication to alleviate symptoms.
- Minimally invasive procedures.
- Surgery to remove or reduce the size of the prostate gland.

Endometriosis

Endometriosis is a medical condition characterized by the growth of uterine lining tissue outside the uterus, resulting in discomfort, inflammation, and possible challenges with fertility. Various strategies to manage endometriosis encompass pain alleviation, hormonal treatments designed to curtail the proliferation of endometrial tissue, and surgical intervention to excise or minimize the impacted tissue.

Erectile Dysfunction (ED)

ED refers to the inability to achieve or maintain an erection sufficient for sexual intercourse. It can have physical or psychological causes. Treatments for ED may involve lifestyle changes, counseling, medication, or medical procedures, depending on the underlying cause.

Menstrual Disorders

Menstrual disorders encompass a range of abnormal menstrual patterns, including heavy or prolonged bleeding (menorrhagia), absent or irregular periods (oligomenorrhea), or severe menstrual pain (dysmenorrhea). Treatments may include hormonal therapies, nonsteroidal anti-inflammatory drugs (NSAIDs) for pain management, or surgical interventions in extreme cases.

These are just a few common reproductive health issues that individuals may encounter. It is essential to seek professional medical advice for accurate diagnosis and personalized treatment options based on the specific condition. Early detection, proper management, and access to reproductive healthcare services are crucial in maintaining reproductive health and well-being.

8

The Integumentary System

T he integumentary system, our body's exterior armor, is an incredible ensemble of skin, hair, nails, and related glands. This system offers physical protection and instrumentalizes temperature balance, sensation, vitamin D synthesis, and immune defense. As we venture into the fascinating realm of the integumentary system, we'll dissect its structure, functions, and astonishing capabilities.

This defensive fortress, the integumentary system, serves as a shield, protecting our bodies from external threats such as pathogens, toxins, and UV radiation. Moreover, it masterfully regulates body temperature through actions like perspiration and the dilation or constriction of blood vessels close to the skin. The skin is further laced with sensory receptors, enabling us to experience sensations like touch, pressure, pain, and temperature variations.

As the body's most extensive organ, the skin is a complex structure composed of three essential layers: the epidermis, dermis, and hypodermis (or subcutaneous tissue). The epidermis, the skin's outermost layer, provides a waterproof barrier and is home to melanocytes that produce melanin, giving our skin its unique color. The dermis, located just beneath the epidermis, hosts an array of blood vessels, hair follicles, sweat glands, and sensory receptors. At its deepest level, the hypodermis is a reservoir of fatty tissue providing insulation and energy storage.

The skin's appendages, hair, and nails, constructed of keratinized cells, offer additional layers of protection and functionality. Hair serves multiple roles, providing insulation, acting as a protective barrier, and aiding sensory perception. Nails found on fingertips and toes protect these sensitive areas while assisting in gripping objects.

Complementing the system are the sebaceous glands, producers of sebum, an oily substance that keeps our skin moisturized and bolsters defense against pathogens. Additionally, sweat glands, including the eccrine and apocrine variants, contribute to temperature regulation and waste excretion. Let's dive deeper into the structure and function of the skin, hair, and nails.

The skin, our protective outer garment, is a complex mosaic of the epidermis, dermis, and hypodermis. Each layer has unique attributes and carries out specific tasks.

The Epidermis

This outer layer of the skin serves as our primary defense against external hazards and UV radiation. It's also a habitat for melanocytes, pigment-producing cells responsible for our skin color, and facilitates skin cell regeneration.

The Dermis

The dermis, nestled beneath the epidermis, provides structural support and nourishment. It's a reservoir of blood vessels, hair follicles, and various glands. It plays a critical role in temperature regulation and sensory perception.

The Hypodermis

Also known as subcutaneous tissue, the hypodermis is the deepest layer of the skin. Composed primarily of fatty tissue, it provides insulation, cushions underlying structures, and stores energy.

Hair, another vital element of the integumentary system, serves various functions, including insulation, protection, and sensory perception, helping us detect environmental changes.

Nails on the fingertips and toes are composed of keratinized cells. They offer protection, assist in handling objects with precision, and, with the help of sensory nerve endings in the nail bed, enable us to detect pressure and touch.

By understanding the structure and functions of the skin, hair, and nails, we gain insight into the remarkable capabilities of the integumentary system. These components work cohesively to protect, regulate, and connect us with our environment, contributing significantly to the well-being of our incredible human bodies.

Common Skin Conditions and Their Treatments

The skin, the outermost protective barrier of our bodies, is susceptible to various conditions and disorders. This section will explore some common skin conditions and the treatments available.

Acne

Acne is a common skin condition characterized by the formation of pimples, blackheads, and whiteheads. It typically occurs during adolescence due to hormonal changes but can affect individuals of any age. Acne treatments include topical creams, cleansers, and oral medications that help reduce inflammation, control oil production, and prevent bacterial growth.

Eczema

Eczema, a chronic inflammatory skin condition, is also known as dermatitis. It results in dry, itchy, and inflamed patches of skin. Moisturizers, corticosteroid creams, and antihistamines are commonly used to manage eczema symptoms and reduce skin inflammation.

Psoriasis

Psoriasis is a chronic autoimmune condition that speeds up the skin cell growth cycle, forming thick, scaly patches on the skin. Topical treatments, phototherapy, systemic medications, and biological therapies alleviate symptoms and manage psoriasis flare-ups.

Dermatitis

Dermatitis is skin inflammation caused by various factors such as irritants, allergens, or genetic predisposition. Treatment involves identifying and avoiding triggers, using corticosteroid creams or ointments, applying moisturizers, and practicing good skincare habits.

Rosacea

Characterized by redness, flushing, and the formation of small blood vessels, rosacea is a chronic inflammatory condition that predominantly affects the face. Treatment options for rosacea include topical medications, oral antibiotics, laser therapy, and lifestyle modifications to manage triggers like sun exposure and certain foods.

Skin Infections

Skin infections, such as bacterial or fungal, can occur due to various factors. Treatment typically involves topical or oral antibiotics or antifungal medications, depending on the specific infection.

Skin Cancer

Skin cancer is a potentially serious condition caused by the abnormal growth of skin cells. Treatment options depend on the type and stage of cancer. Still, they may include surgical removal, radiation therapy, chemotherapy, or targeted therapy.

It is important to note that proper diagnosis and treatment should be sought from qualified healthcare professionals or dermatologists for these and other skin conditions. The treatment approach will depend on the situation, severity, and individual factors.

Understanding common skin conditions and their available treatments enables us to actively care for our skin health, seek timely interventions, and promote overall well-being. Remember, healthy and radiant skin contributes not only to our physical appearance but also to our confidence and self-esteem.

9

The Endocrine System

Embarking on a fascinating exploration of our body's internal communication network, we delve into the realm of the endocrine system. This complex maze of glands, responsible for producing and disseminating hormones, masterfully choreographs various vital bodily functions.

The endocrine system deftly ensures homeostasis, guiding numerous physiological processes. It deftly modulates our growth, metabolism, reproduction, mood, and overall function. With this robust network, our bodies would gain the direction to maintain optimal health.

Let's discover more about the endocrine system's maestros - the glands. Scattered throughout our body, they form a stellar cast, including the pituitary, thyroid, adrenal glands, pancreas, parathyroid, pineal gland, ovaries (in females), and testes (in males). Each gland meticulously produces specific hormones, their roles defined by their unique effects on target organs and tissues.

Hormones, the endocrine system's secret messages, traverse through our bloodstream to their destination cells or organs. They bind to specific receptors on reaching, setting off a cascade of physiological responses. These potent substances influence many functions, including metabolism, growth, development, reproduction, stress responses, blood sugar regulation, and electrolyte balance.

The hypothalamus, a humble region in the brain, creates a crucial bridge between the nervous and endocrine systems. It directs the pituitary gland, often celebrated as the 'master gland.' The pituitary, in turn, regulates growth and sexual development and releases a gamut of hormones controlling various body processes.

The thyroid and parathyroid glands regulate our metabolism, energy production, body temperature, and blood calcium levels. Meanwhile, our adrenal glands, perched atop our kidneys, manage stress response, metabolism, salt and water balance, and blood pressure regulation.

The pancreas plays dual roles as an endocrine and exocrine gland, ensuring blood sugar levels and energy storage balance. Any malfunction in this crucial gland can result in diabetes mellitus, impaired insulin production, or insulin resistance.

With an enriched understanding of the endocrine system, we'll truly appreciate its pivotal role in choreographing our growth, metabolism, reproduction, and the delicate dance of our hormonal balance. So, hop on board for this enlightening expedition as we navigate the winding trails and extraordinary influences of hormones on our bodies and minds.

Regulation of Bodily Functions through Hormone Signaling

Hormones are vital in regulating diverse physiological functions and promoting overall homeostasis. This section will explore how hormone signaling regulates different physiological processes throughout the body.

Hormone Signaling

Hormones are chemical messengers released by endocrine glands, traveling through the bloodstream. Once in the bloodstream, they reach target cells or tissues, binding to specific receptors. This binding triggers a cascade of biochemical reactions within the cells, ultimately leading to specific physiological responses.

Feedback Mechanisms

The regulation of hormone secretion is tightly controlled by feedback mechanisms to maintain balance within the body. Two primary types of feedback mechanisms exist:

- **Negative Feedback:** When hormone levels reach a certain threshold, negative feedback mechanisms inhibit further hormone secretion. This helps prevent excessive hormone production and maintains stability within the body.
- **Positive Feedback:** Positive feedback mechanisms sometimes amplify a physiological response. This occurs when the initial stimulus promotes increased hormone secretion, further enhancing the reaction.

Hormonal Regulation of Bodily Functions

The endocrine system, through its intricate network of hormones, influences a wide range of bodily functions, including:

- **Metabolism:** Hormones such as insulin, glucagon, thyroid hormones, and cortisol regulate metabolism by controlling energy production, glucose utilization, and lipid metabolism.
- **Growth and Development:** Growth hormones, thyroid hormones, and sex hormones play critical roles in regulating growth, development, and sexual maturation.
- **Reproduction:** Hormones like estrogen, progesterone, testosterone,

follicle-stimulating hormone (FSH), and luteinizing hormone (LH) regulate reproductive processes, including fertility, menstruation, and pregnancy.

- **Stress Response:** Hormones like cortisol and adrenaline (epinephrine) are involved in the body's response to stress, enabling it to cope with challenging situations.
- **Calcium Balance:** Parathyroid hormone (PTH) and calcitonin regulate calcium levels in the blood, ensuring proper bone health and nerve function.
- **Water and Electrolyte Balance:** Hormones such as aldosterone and antidiuretic hormone (ADH) help maintain the body's water and electrolytes balance by influencing kidney function.

Imbalances and Disorders of the Endocrine System

While the endocrine system maintains balance and proper functioning within the body, imbalances and disorders can disrupt this delicate equilibrium. This section will explore common imbalances and disorders of the endocrine system and their effects on overall health.

Hormone Imbalances

Hormone imbalances can occur when there is either an excess or deficiency of specific hormones in the body. These imbalances can arise from various factors, including genetic predispositions, lifestyle choices, environmental factors, or underlying medical conditions. Some common hormone imbalances include:

- **Hyperthyroidism:** An overactive thyroid gland leading to excessive production of thyroid hormones, resulting in symptoms such as weight loss, rapid heartbeat, and anxiety.
- **Hypothyroidism:** An underactive thyroid gland leading to insufficient

production of thyroid hormones, causing symptoms such as fatigue, weight gain, and depression.

- **Diabetes:** A disorder characterized by high blood sugar levels due to either insufficient insulin production (Type 1 diabetes) or impaired insulin function (Type 2 diabetes).
- **Adrenal Insufficiency:** Occurs when the adrenal glands do not produce enough cortisol and aldosterone, leading to fatigue, low blood pressure, and electrolyte imbalances.
- **Pituitary Disorders:** Abnormalities in the pituitary gland can cause disruptions in various hormone production, leading to conditions like growth hormone deficiency or pituitary tumors.
- **Polycystic Ovary Syndrome (PCOS):** A hormonal disorder in females that can cause irregular menstrual cycles, excessive hair growth, and fertility issues.

Endocrine Disorders

Apart from hormone imbalances, specific disorders affect the endocrine system. These disorders can impact the function of multiple glands or disrupt hormone production and regulation. Some examples include:

- **Cushing's Syndrome:** A condition characterized by excessive cortisol levels in the body, leading to weight gain, muscle weakness, and mood changes.
- **Addison's Disease:** Manifests when the adrenal glands fail to generate sufficient amounts of cortisol and aldosterone, leading to symptoms such as fatigue, low blood pressure, and electrolyte imbalance.
- **Hypopituitarism:** A condition where the pituitary gland fails to produce one or more hormones, causing a range of symptoms depending on the affected hormones.

Impact on Health and Treatment

Imbalances and disorders of the endocrine system can significantly affect overall health and well-being. They can disrupt metabolism, growth and development, reproductive functions, and other physiological processes. Proper diagnosis and treatment are crucial in managing these conditions. Depending on the disorder's severity, treatment options may include hormone replacement therapy, lifestyle modifications, medication, or surgery.

Understanding the imbalances and disorders within the endocrine system is vital in recognizing symptoms, seeking appropriate medical care, and maintaining optimal health.

10

The Immune System

Our bodies are home to a fascinating defense network - the immune system - consisting of organs, cells, and molecules that collaborate harmoniously to protect us from harmful invaders like bacteria, viruses, and parasites. Prepare yourself for a captivating exploration of this labyrinth, where we decode its critical functions in safeguarding our health and maintaining our well-being at its zenith.

Think of the immune system as an expert military defense unit with two primary strategies: the innate and the adaptive immune system. The inherent division serves as our body's rapid, broad-spectrum defense force. In contrast, the adaptive division develops targeted, long-lasting immunity against specific pathogens.

White blood cells - the soldiers of our immune army - come in various types, such as neutrophils, lymphocytes, and macrophages, each playing their part in detecting and eliminating pathogens. Lymphocytes, including T and B cells, are the linchpins of our adaptive immune responses.

But what's an army without a fortress? The immune system's stronghold is an assortment of lymphoid organs like the thymus, spleen, and lymph nodes, where immune cells mature and prepare for action. Other vital components include antibodies - specialized proteins that mark specific pathogens for destruction - and cytokines, the communication channels that regulate immune responses.

The immune system responds to invading pathogens like a well-coordinated military operation. It starts with pathogen recognition and immune cell activation, progresses to inflammation, and culminates in a specific T and B-cell-mediated response. This response doesn't just fight the present invasion; it also prepares us for future encounters with the same enemy – a remarkable quality we leverage in vaccinations to provide long-term immunity.

However, like all things complex, the immune system is full of challenges. Autoimmune diseases occur when it mistakenly attacks our tissues, while immunodeficiency disorders weaken the immune response, making us more susceptible to infections. We can support and maintain a robust immune system through mindful practices like proper nutrition, adequate sleep, stress management, and healthy lifestyle choices.

In this chapter, we'll dive deeper into the complex labyrinth of the immune system. We'll explore the fundamental mechanisms, components, and astounding ways this biological marvel keeps us safe from diseases. Additionally, we'll discuss the exciting domain of immunology to reveal breakthroughs in research and a glimpse at the future of immune-based therapies and treatments.

Role of the Immune System in Defending Against Pathogens

Innate Immunity

The first line of defense is the innate immune response, which provides immediate, nonspecific protection against various pathogens. Critical components of innate immunity include:

- **Physical Barriers:** The skin and mucous membranes act as physical barriers, preventing pathogens from entering the body. They secrete substances that inhibit the growth of microorganisms.
- **Phagocytes:** Phagocytes, including neutrophils and macrophages, engulf and destroy pathogens through phagocytosis.
- **Natural Killer Cells:** Natural killer cells detect and destroy virus-infected cells and cancerous cells, helping to prevent the spread of infection.
- **Inflammatory Response:** Inflammation is a localized reaction triggered by tissue injury or infection, characterized by heightened blood flow, the recruitment of immune cells, and the release of inflammatory mediators. These processes collectively aim to eliminate pathogens and facilitate the healing process.

Adaptive Immunity

The adaptive immune response is a highly specialized defense mechanism that develops over time and provides long-term immunity against specific pathogens. It involves two key components:

- **B Cells and Antibodies:** B cells produce antibodies that recognize and bind to specific antigens in pathogens. These antibodies neutralize the pathogens, mark them for destruction, and activate other immune cells to eliminate them.
- **T Cells:** T cells assume a vital role in cell-mediated immunity. They can directly recognize and eliminate infected cells, stimulate B cells to produce antibodies or regulate the immune response.

Immune Memory

One of the most remarkable features of the immune system is its ability to develop immunological memory. Upon encountering a pathogen for the first time, the immune system mounts a primary immune response. Subsequent encounters with the same pathogen lead to a faster and stronger secondary immune response due to the presence of memory B and T cells. This memory response provides long-lasting protection against reinfection.

Immunization

Immunization, or vaccination, is a powerful tool that harnesses the immune system's ability to develop memory responses. Vaccines contain harmless antigens that stimulate an immune response, allowing the body to recognize and remember the pathogen. This way, if the individual encounters the pathogen in the future, their immune system can mount a rapid and effective response, preventing disease development.

Autoimmune Diseases and Immune System Disorders

While the immune system's primary function is to shield the body from external threats, there are instances when it may malfunction and erroneously target its cells and tissues. These conditions are known as autoimmune diseases and immune system disorders, and they can profoundly impact an individual's health and well-being. This section will explore some common autoimmune diseases and immune system disorders.

Autoimmune Diseases

Autoimmune diseases occur when the immune system mistakenly identifies normal, healthy cells as foreign and launches an immune response against them. Some well-known autoimmune disorders include:

- **Rheumatoid Arthritis:** This chronic inflammatory disease primarily affects the joints, causing pain, swelling, and stiffness.
- **Systemic Lupus Erythematosus:** Lupus is a systemic autoimmune

disease that can affect multiple organs and tissues, leading to symptoms such as fatigue, joint pain, skin rashes, and kidney problems.

- **Multiple Sclerosis:** In this autoimmune condition, the immune system launches an assault on the protective coating of nerve fibers in the central nervous system, causing disruptions in the communication between the brain and the rest of the body.
- **Type 1 Diabetes:** Type 1 diabetes arises when the immune system erroneously targets and destroys the insulin-producing cells within the pancreas, leading to elevated blood sugar levels.

Allergies

Allergies are hypersensitivity reactions triggered by exposure to harmless substances, such as pollen, dust mites, or certain foods. In individuals with allergies, the immune system overreacts to these substances, leading to symptoms like sneezing, itching, hives, and in severe cases, anaphylaxis.

Immunodeficiency Disorders

Immunodeficiency disorders occur when the immune system is weakened or compromised, leaving the body susceptible to infections. Some examples of immunodeficiency disorders include:

- **Primary Immunodeficiency Disorders:** These genetic conditions result in a weakened immune system from birth, making individuals more susceptible to infections.
- **Acquired Immunodeficiency Syndrome (AIDS):** AIDS is the consequence of infection by the human immunodeficiency virus (HIV), which targets and eradicates the CD4 cells of the immune system, leaving the body susceptible to opportunistic infections.

Hypersensitivity Reactions

Hypersensitivity reactions are exaggerated immune responses to specific substances. The immune system reacts as if the substance is harmful, triggering an immune response that can cause symptoms ranging from mild to severe. Examples include allergic rhinitis (hay fever), asthma, and drug allergies.

Autoinflammatory Disorders

Autoinflammatory disorders are characterized by recurrent episodes of inflammation in the absence of an autoimmune or infectious cause. These conditions result from a dysfunction in the innate immune system, leading to uncontrolled inflammation. Examples include familial Mediterranean fever and periodic fever syndromes.

Understanding autoimmune diseases and immune system disorders is crucial for early detection, diagnosis, and management. The continuous progress in research and treatment options provides optimism for individuals grappling with these conditions, striving to enhance their quality of life and overall well-being.

11

The Urinary System

The urinary system, also known as the renal system, plays a vital role in maintaining the body's internal balance by regulating fluid levels, electrolytes, and waste elimination. Comprised of several organs working together, the urinary system ensures the proper functioning of the body's filtration and excretion processes.

Overview of the Urinary System

Our bodies are complex and elaborate systems with myriad subsystems collaborating seamlessly. Among these, the urinary system stands out as an impressive biological marvel. This essential network, composed of the kidneys, ureters, bladder, and urethra, tirelessly purges waste from our bodies and regulates our water balance, ensuring optimal health and hydration.

At the heart of this system lie the kidneys, our sophisticated filtration and reabsorption centers. The kidneys, nestled within the upper abdominal cavity, are bean-shaped powerhouses that contain millions of tiny nephrons, which are the building blocks of the filtration process. Beyond this pivotal task, the kidneys balance fluid and electrolyte levels, stabilize blood pressure, and produce hormones that manage red blood cell production and calcium metabolism.

The fascinating journey of urine formation commences in the nephrons. As blood courses into the kidneys through the renal arteries, the nephrons meticulously sieve out waste, excess water, and electrolytes. The resulting primary urine still contains nutrients and water, diligently reabsorbed into the bloodstream, maintaining the body's delicate equilibrium. The remaining purified urine then journeys to the bladder, a reservoir awaiting expulsion, marking the end of a sophisticated cycle of filtration and regulation.

Now, let's delve deeper into each component of the urinary system, starting with the kidneys. Located on either side of the spine in the upper abdomen, they are the unsung heroes regulating our body's water balance, electrolytes, and acids while helping control blood pressure and stimulating red blood cell production.

Continuing our exploration, we encounter the ureters, which function as slender ducts that act as conduits, transporting urine from the kidneys to the bladder. To ensure a one-way flow, the smooth muscle walls of the ureters undergo rhythmic contractions.

The bladder, a muscular sac behind the pubic bone in the lower abdomen, provides temporary storage for urine. This unique structure, lined with transitional epithelium, expands and contracts to accommodate varying volumes of urine, preventing leakage.

Lastly, the urethra is the final pathway that conveys the urine from the bladder out of the body during urination. In males, this tube also serves as a conduit for semen during ejaculation. Its length varies between the sexes, being typically longer in males.

Collectively, the kidneys, ureters, bladder, and urethra form a harmonious ensemble, ensuring the urinary system's proper functioning. The kidneys filter waste and excess fluids from the blood, yielding urine that the ureters channel to the bladder. This temporary storage vessel holds the urine until it's voluntarily excreted during urination.

The urinary system plays a crucial role in waste elimination and maintaining the body's fluid, electrolyte, and acid-base balance. The kidneys keep electrolyte concentrations, including sodium, potassium, and calcium, within a healthy range, essential for optimal physiological functioning.

Post-filtration, the renal tubules reclaim vital substances like water, glucose, and electrolytes, transporting them back into the bloodstream through surrounding capillaries. Simultaneously, additional waste is secreted into the tubules. This selective process preserves essential substances while maintaining fluid balance.

Further aiding balance, the kidneys regulate urine concentration and dilution via hormonal signals that adjust renal tubule permeability. When the body needs to conserve water, the tubules reabsorb more, resulting in concentrated urine. Conversely, when water needs to be expelled, the tubules reabsorb less, diluting the urine.

This complex yet efficient filtration and waste elimination process is fundamental in preserving the body's internal balance and expelling harmful substances, ensuring our body functions properly. In the upcoming sections, we'll explore common disorders and diseases affecting the urinary system, along with the available diagnostic techniques and treatment options.

Common Urinary System Disorders and Their Treatments

The urinary system can be susceptible to various disorders and conditions affecting normal functioning. This section will explore some common urinary system disorders and the available treatments for each.

Urinary Tract Infections (UTIs)

UTIs occur when bacteria enter the urinary tract and multiply, leading to infection. The most common UTIs affect the bladder (cystitis) or the urethra (urethritis). Possible symptoms include increased urination frequency, burning sensation while urinating, cloudy or bloody urine, and pelvic pain. UTIs are typically treated with antibiotics to eliminate the bacterial infection. Drinking plenty of water and practicing good hygiene can help prevent UTIs.

Kidney Stones

Kidney stones are hard mineral and salt deposits that form in the kidneys. As they traverse the urinary tract, they can induce intense pain. Treatment options for kidney stones depend on their size and location. Small stones may pass naturally with increased fluid intake and pain management. Larger stones may require medical intervention, such as extracorporeal shock wave lithotripsy (ESWL), ureteroscopy, or surgical removal.

Urinary Incontinence

Urinary incontinence is characterized by the unintentional release of urine. It can result from weakened pelvic floor muscles, nerve damage, or underlying medical conditions. Treatment options for urinary incontinence include pelvic floor exercises (Kegel exercises), bladder training techniques, medication, and in severe cases, surgery. Making lifestyle adjustments, such as steering clear of bladder irritants and maintaining a healthy weight, can also aid in managing the condition.

Urinary Retention

Urinary retention occurs when the bladder cannot empty completely, leading to a persistent feeling of incomplete bladder emptying. It can be caused by various factors, including bladder muscle dysfunction, obstruction, or nerve damage. Treatment options for urinary retention depend on the underlying cause. They may include medication to relax the bladder muscles, catheterization to empty the bladder, or surgical procedures to address structural issues.

Bladder Cancer

Bladder cancer refers to the abnormal growth of cells in the bladder lining. Symptoms may include blood in the urine, frequent urination, and pain during urination. Treatment options for bladder cancer depend on the stage and grade of the tumor. They may involve surgery, radiation therapy, chemotherapy, or immunotherapy. Early detection and prompt treatment are crucial for better outcomes.

Urinary System Infections and Inflammation

Apart from UTIs, other infections and inflammatory conditions can affect the urinary system, such as interstitial cystitis, prostatitis, or pyelonephritis. Treatment approaches may include antibiotics, anti-inflammatory medications, pain management, and lifestyle modifications to alleviate symptoms and manage the underlying causes.

It's important to note that each urinary system disorder is unique, and treatment plans should be tailored to the individual's specific condition and medical history. Suppose you experience any symptoms or concerns related to the urinary system. In that case, it is advisable to consult a healthcare professional for an accurate diagnosis and appropriate treatment.

12

Mind-Blowing Facts and Discoveries

T he human body continues to be a subject of awe and fascination, and scientific research has led to numerous remarkable discoveries and advancements. This chapter will delve into some impressive facts and breakthroughs that have revolutionized our understanding of the human body.

Genomic Sequencing

Completing the Human Genome Project in 2003 was a monumental achievement that provided a comprehensive map of the human genetic code. This breakthrough has enabled scientists to study the relationships between genes, diseases, and inherited traits, opening doors to personalized medicine and targeted therapies.

Neuroplasticity

The discovery of neuroplasticity shattered the long-held belief that the brain is a fixed and unchanging organ. Research has shown that the brain can adapt and rewire itself, forming new neural connections and pathways throughout life. This finding has profound implications for brain rehabilitation, learning, and understanding various mental health conditions.

Gut-Brain Connection

Emerging research has revealed the relationship between the gut and the brain, known as the gut-brain axis. The gut houses trillions of microorganisms, collectively known as the gut microbiota, which play a crucial role in digestion and metabolism. Moreover, these microbes communicate with the brain through complex pathways, influencing mood, cognition, and even behavior. Understanding this connection opens up new avenues for treating mental health disorders and gastrointestinal conditions.

Regenerative Medicine

Advancements in regenerative medicine have shown promise in repairing and replacing damaged tissues and organs. Stem cell research, tissue engineering, and 3D printing technologies have paved the way for regenerating tissues such as skin, cartilage, and organs like the liver and heart. These breakthroughs offer hope for patients with debilitating injuries or organ failure.

CRISPR-Cas9 Gene Editing

The development of CRISPR-Cas9 gene editing technology has revolutionized the field of genetics. This powerful tool allows scientists to precisely edit genes, opening up possibilities for treating genetic disorders, modifying traits, and potentially curing diseases. CRISPR can potentially reshape the future of medicine and has garnered significant attention in the scientific community.

Microbiome Research

The human microbiome, consisting of the vast community of microorganisms inhabiting our bodies, has emerged as a fascinating study area. Research has shown that these microbes influence our health, metabolism, immune system, and mental well-being. Understanding the complexity of the microbiome has the potential to transform healthcare practices and lead to innovative treatments.

Artificial Intelligence in Healthcare

Artificial intelligence (AI) has made significant strides in healthcare, aiding diagnosis, treatment planning, and data analysis. Machine learning algorithms and AI systems can process vast amounts of medical data, leading to more accurate diagnoses, personalized treatment recommendations, and improved patient outcomes.

These remarkable discoveries and advancements are just a glimpse of the ongoing research and exploration of the human body. As scientists continue to uncover the mysteries of our spectacular biology, the potential for new breakthroughs and transformative discoveries is limitless.

Surprising Facts and Mysteries about the Human Body

The human body is a remarkable creation full of astonishing features and mysteries yet to be fully understood. This chapter will explore some surprising facts and delve into the enigmatic aspects that continue to captivate scientists and researchers worldwide.

Superhuman Strength

Under extreme circumstances, individuals have displayed incredible feats of strength. Adrenaline, the "fight or flight" hormone, can temporarily boost strength and endurance, allowing individuals to lift heavy objects or perform remarkable physical tasks. The extent of this hidden strength potential is still a subject of fascination and exploration.

Phantom Limb Sensation

People who have undergone amputation sometimes experience the sensation that their missing limb is still present. This phenomenon, known as phantom limb sensation, is believed to be caused by the brain's inability to adjust to the absence of the limb fully. The brain continues receiving signals from the nerves previously connected to the amputated limb, creating a vivid and perplexing experience.

Unique Fingerprints

No two individuals have the same fingerprints. Even identical twins share the same DNA and have distinct fingerprint patterns. The ridges and loops on our fingertips are formed during fetal development and remain unchanged throughout our lives. The exact reason behind the uniqueness of fingerprints remains a fascinating mystery.

Sleep Paralysis

Sleep paralysis is when a person is temporarily unable to move or speak while falling asleep or waking up. It occurs due to a disruption in the transition between sleep stages, leaving individuals briefly conscious but unable to control their muscles. This intriguing phenomenon has sparked various cultural interpretations and has been linked to paranormal experiences and hallucinations.

Handedness

Approximately 90% of people are right-handed, while 10% are left-handed or ambidextrous. The reasons behind handedness still need to be fully understood. Still, genetic, environmental, and cultural factors are believed to influence it. The dominance of one hand over the other adds an intriguing dimension to the complexity of human behavior.

Biological Clocks

The human body operates on various internal biological clocks that regulate essential functions such as sleep-wake cycles, hormone production, and body temperature. The body's internal timekeepers, called circadian rhythms or biological clocks, are impacted by external elements like light and hold significant importance in sustaining overall health and well-being. The intricacies of these internal timekeepers continue to be an area of active research.

Human Longevity

While the human lifespan has limitations, researchers continually explore the factors that contribute to exceptional longevity. Certain regions of the world, known as Blue Zones, have a higher concentration of individuals living beyond 100 years. Genetics, lifestyle choices, and environmental factors are believed to play a role in these remarkable cases, providing insights into the possibilities of extending human life.

Cutting-Edge Technologies and Techniques Used in Studying the Human Body

Advancements in technology and innovative research techniques have dramatically enhanced the exploration of the human body. This section will explore the cutting-edge tools and methodologies that have revolutionized our understanding of the human body.

Magnetic Resonance Imaging (MRI)

MRI technology utilizes powerful magnetic fields and radio waves to generate detailed images of the body's internal structures. It provides a non-invasive and highly accurate way to visualize organs, tissues, and brain. MRI has become a cornerstone of medical diagnostics and research, enabling us to examine the human body in unprecedented detail.

Genomic Sequencing

The Human Genome Project marked a significant milestone in our understanding of human genetics. Today, genomic sequencing techniques have become more advanced and affordable, allowing scientists to decode an individual's entire genetic blueprint. This breakthrough technology has opened up new avenues for personalized medicine, disease research, and the exploration of our genetic heritage.

Electron Microscopy

Electron microscopy enables scientists to visualize structures at an incredibly high resolution. Researchers can examine tissues, cells, and even individual molecules with extraordinary precision using a beam of accelerated electrons. This technique has unveiled details of cellular structures and processes, providing invaluable insights into the complexity of the human body.

Optogenetics

Optogenetics combines optics and genetics to manipulate and observe the activity of specific cells in living organisms. By introducing light-sensitive proteins into targeted cells, researchers can control their function with precise timing. This powerful technique has shed light on the complex workings of the brain. It holds promise for understanding neurological disorders and developing new therapeutic approaches.

Functional Magnetic Resonance Imaging (fMRI)

fMRI measures changes in blood flow within the brain to map brain activity and identify functional areas associated with specific tasks or stimuli. It provides insights into how different brain regions are interconnected and work together. fMRI has been instrumental in discovering the mysteries of cognition, emotion, and neurological disorders.

Single-Cell Analysis

Advancements in single-cell analysis techniques have enabled scientists to study individual cells with unprecedented detail. By analyzing a single cell's genetic, molecular, and functional characteristics, researchers can uncover cellular diversity, understand cell development, and study disease mechanisms at a cellular level. This approach has the potential to revolutionize our understanding of complex biological systems.

These cutting-edge technologies and techniques are propelling human body research into new frontiers. They empower scientists to unravel the complexities of our anatomy, physiology, and genetic makeup, leading to breakthrough discoveries and transformative advancements in medicine and healthcare.

Conclusion

Throughout this captivating exploration of the human body in "The Abyss Inside: Mind-Blowing Facts and Discoveries About Your Extraordinary Human Body," we have embarked on an awe-inspiring journey of discovery. From the compounded systems that sustain our existence to the wondrous mechanisms that enable our every thought, movement, and sensation, we have explored the depths of our miraculous biology. As we conclude this enlightening journey, let us recap the key points and insights we have gained.

Anatomy

We have unraveled the marvels of human anatomy, comprehending the structure, organization, and interconnectedness of our bones, muscles, nerves, and organs. The human body is a detailed masterpiece, with each component crucial in maintaining our well-being.

Circulatory System

We have marveled at the circulatory system, the vital transportation network that delivers oxygen, nutrients, and essential substances to every cell and removes waste products. We have explored the pulmonary and systemic circuits, gaining an understanding of how blood flows and its crucial role in maintaining equilibrium.

Digestive System

We have embarked on a journey through the digestive system, witnessing the transformation of food into nourishment. We have discovered the significance of each organ, from the mouth to the intestines, and the remarkable digestion and nutrient absorption process.

Nervous System

We have unraveled the mysteries of the nervous system, the complex network that enables communication and coordination throughout our body. We have explored the structure and function of neurons and glial cells, delving into the central and peripheral nervous systems and their roles in sensory perception and motor control.

Reproductive System

We have explored the wonders of the reproductive system, understanding the unique processes of male and female anatomy, puberty, menstruation, and fertility. We have contemplated the miraculous creation of life and the balance of hormones that orchestrate reproduction.

Integumentary System

We have appreciated the remarkable functions of the integumentary system, including the skin, hair, and nails, which provide protection, regulation, and sensory perception. We have learned about the skin's role as a sensory interface with the world and the significance of its health and care.

Endocrine System

We have delved into the endocrine system, unraveling the vital role of hormones in regulating bodily functions. We have explored the major endocrine glands and their profound impact on growth, metabolism, reproduction, and overall well-being.

Immune System

We have marveled at the immune system, our impressive defense against pathogens and foreign invaders. We have learned about immune response mechanisms and the significance of white blood cells and lymph nodes in maintaining our health.

Urinary System

We have discovered the remarkable role of the urinary system in maintaining fluid balance, filtering waste products, and regulating blood pressure. We have explored the anatomy and functions of the kidneys, bladder, ureters, and urethra.

Fascinating Discoveries

The fascinating discoveries and advancements in human body research have enthralled us. Throughout our journey, we have encountered astonishing revelations, enigmatic phenomena, and state-of-the-art technologies that consistently challenge the limits of our comprehension.

"The Abyss Inside: Mind-Blowing Facts and Discoveries About Your Extraordinary Human Body" has unveiled our mysteries, awakening a deep sense of wonder and appreciation for the incredible vessel that houses our consciousness. May this knowledge inspire us to nurture and care for our bodies, celebrate their resilience and adaptability, and embark on a lifelong journey of exploration and self-discovery.

Encouragement to Continue Exploring and Appreciating the Wonders of the Human Body

As we reach the end of our journey through "The Abyss Inside: Mind-Blowing Facts and Discoveries About Your Extraordinary Human Body," let us not view this as the culmination of our exploration but rather as a stepping stone towards a lifelong appreciation and fascination with the wonders of the human body.

The knowledge we have acquired about our anatomy, physiology, and interwoven systems is just the beginning. It is an invitation to dive deeper, to continue seeking knowledge and understanding about the complexities that make us who we are.

Every day, remarkable discoveries and advancements are being made in human biology. New insights emerge, unraveling further mysteries and expanding our understanding of the intricacies of our existence. Embrace this spirit of curiosity and embark on a personal journey of exploration.

Take the time to listen to your body, to be attuned to its signals and needs. Recognize the remarkable synchronicity between your thoughts, emotions, and physical sensations. Cherish the awe-inspiring moments when you witness the body's resilience, adaptability, and capacity for healing.

Engage in conversations with experts, researchers, and fellow enthusiasts who share your passion for the human body. Attend lectures, read books, and explore online resources to stay abreast of the latest discoveries and developments. Seek out opportunities to participate in studies or contribute to ongoing research projects.

Engaging in physical activities, such as exercise, dance, or yoga, allows us to connect with our bodies profoundly. By nurturing our physical well-being, we foster a deeper understanding and appreciation for the marvels that lie within.

Moreover, let us recognize the interconnectedness of our bodies with the natural world. Appreciate the role of nutrition, rest, and environmental factors in maintaining optimal health. Embrace a holistic approach that recognizes the synergy between our bodies, minds, and the world we inhabit.

As we conclude our exploration of the human body, remember that the journey does not end here. It is an ongoing quest, an opportunity to deepen our understanding and a lifelong celebration of the magnificent vessel that carries us through life.

May the knowledge and insights gained from "The Abyss Inside: Mind-Blowing Facts and Discoveries About Your Extraordinary Human Body" inspire you to continue exploring, appreciating, and nurturing the wonders of your own body. Embrace the gift of self-discovery and let it fuel your curiosity, awe, and reverence for the magnificent creation of the human body.

Help Guide Future Explorers: Share Your Thoughts

Thank you for journeying inward with us in "The Abyss Inside: Mind-Blowing Facts and Discoveries About Your Extraordinary Human Body." We trust you've gained fascinating insights and deepened your appreciation for the marvel that is the human body.

As your journey ends, we invite you to share your reflections on this adventure. By leaving a review on Amazon, you help us continue to improve and tailor our content and provide guidance to other explorers embarking on similar journeys of self-discovery. Were there any particular facts or discoveries that amazed you? How has this exploration altered your perception of the human body?

Leaving a review is simple. Visit the book's Amazon page, find the 'Customer Reviews' section, and click 'Write a customer review.' Your insights can help illuminate this path of exploration for others.

Thank you for your continued support and for choosing "The Abyss Inside" as your guide in this extraordinary journey of discovery.

Resources

AAFP. (n.d.). American Academy of Family Physicians. http://www.aafp.org/ *Access anytime anywhere | Cleveland Clinic*. (n.d.). Cleveland Clinic. https://my.clevelandclinic.org/

American Heart Association | To be a relentless force for a world of longer, healthier lives. (n.d.). www.heart.org. http://www.heart.org/ *CDC works 24/7*. (2023, July 19). Centers for Disease Control and Prevention. http://www.cdc.gov/

Fish, T. (2021, October 27). 12 Mind-Blowing Facts about your Body. *Newsweek*. https://www.newsweek.com/mind-blowing-facts-about-your-body-human-1638872

Home. (2023, July 18). http://www.who.int/ https://website-designer-2149.business.site/. (n.d.). *WEIGHT-CUTTING, PEDs, AND TRT - The Ultimate Guide to Preventing and Treating MMA Injuries: Featuring advice from UFC Hall of Famers Randy Couture, Ken Shamrock, Bas Rutten, Pat Miletich, Dan Severn and more! (2016)*. Publicism - Non-fiction Documentary Books. https://publicism.info/sports/mma/9.html

Information and Resources about for Cancer: Breast, Colon, Lung, Prostate, Skin. (n.d.). American Cancer Society. http://www.cancer.org/

Interesting facts about the human body | SelectHealth. (2020, February 15). SelectHealth.org. https://selecthealth.org/blog/2020/02/15-interesting-facts-about-the-human-body

MedlinePlus. (n.d.). *MedlinePlus - Health Information from the National Library of Medicine.* https://medlineplus.gov/

National Institutes of Health (NIH). (n.d.). National Institutes of Health (NIH). http://www.nih.gov/

Neurological disorders. (2023, May 3). Johns Hopkins Medicine. https://www.hopkinsmedicine.org/health/conditions-and-diseases/neurological-disorders

Ng, A. (2023, June 29). *15 Facts About The Human Body! - National Geographic Kids.* National Geographic Kids. https://www.natgeokids.com/uk/discover/science/general-science/15-facts-about-the-human-body/

No. 1 Hospital in the Nation – Mayo Clinic. (n.d.). Mayo Clinic. http://www.mayoclinic.org/

Pjallens. (2023, January 19). Paul Presents The Berkshire MS Therapy Centre with Kim Williams | West Berkshire Villagers. *West Berkshire Villagers.* https://westberksvillagers.com/paul-presents-the-berkshire-ms-therapy-centre-with-kim-williams/

Top 10 diseases of the nervous system. (2023, March 15). https://db4phone.com/en/post/top-10-diseases-of-the-nervous-system.html

Tpt, T. P. (n.d.). *Roshiita | Nephrology and Urology I Book now Kidney-and-urinary-tract-diseases - in Algeria.* https://roshiita.com/en/Algeria/Blog/Kidney-and-urinary-tract-diseases

Vanstone, E. (2022). Human body Facts. *Science Experiments for Kids.* https://www.science-sparks.com/human-body-facts/

WebMD - Better information. Better health. (2023, July 19). WebMD. http://ww

w.webmd.com/

What Are The Signs of Central Precocious Puberty (CPP)? (n.d.). https://www.pu
bertytoosoon.com/about-cpp/what-are-the-signs-of-central-precocious-pu
berty

Wikipedia contributors. (2023d). Human body. *Wikipedia.* https://en.wikipe
dia.org/wiki/Human_body

www.ingramcontent.com/pod-product-compliance
Lightning Source LLC
Chambersburg PA
CBHW071211120626
46546CB00006B/2507